Clinical Biochemistry

Clinical Biochemistry
A Laboratory Guide

Rooma Devi | Aman Chauhan
Simmi Kharb | Chandra Shekhar Pundir

Published by

Jenny Stanford Publishing Pte. Ltd.
101 Thomson Road
#06-01, United Square
Singapore 307591

Email: editorial@jennystanford.com
Web: www.jennystanford.com

British Library Cataloguing-in-Publication Data
A catalogue record for this book is available from the British Library.

Clinical Biochemistry: A Laboratory Guide

ISBN 978-981-4968-75-1 (Hardcover)
ISBN 978-1-003-45566-0 (eBook)

Contents

Preface

A knowledge of the practical aspects of clinical biochemistry is essential for medical students to understand the diagnostic and prognostic status of diseases. We have always felt the need for a complete and comprehensive book for orientation toward the various aspects of practical biochemistry and this book has been developed to fulfill the same purpose. One of the main objectives of this book is to assist students in their research by providing them with sufficient background information so that they can design experiments and identify the methods best suited to address their specific research questions or problems. The book addresses and compiles some of the invariably used protocols in biochemistry for easy access of the researchers and students, with the aim to help them understand some of the basics behind following the methodologies, including those for urine analysis; carbohydrate, proteins, and lipid analysis; and diabetic and lipid profiles. It also discusses organ function tests, such as liver, renal, pancreatic, and cardiac function tests, in addition to exploring pH and buffer, colorimeter and spectrophotometer, chromatography, electrophoresis, automation, flow cytometry, enzyme-linked immunosorbent assay (ELISA), radioimmunoassay, polymerase chain reaction, and DNA isolation techniques. The book also focuses on the lab safety rules and laboratory glassware. It also provides case histories and multiple choice questions to help students boost their knowledge as well as helps them understand the quality management system, which is an important aspect to obtain accurate laboratory results. It describes and characterizes different types of transducers that are used in the construction of biosensors/ glucometers, which are used to detect glycated hemoglobin, glycated albumin, fractosamine, etc.

The book is well-suited for undergraduate, graduate, postgraduate, and professional courses, covering the practical syllabus for MBBS, MD, BDS, BSc MLT, BSc nursing, and MSc students, as it will help students acquire skills to work in clinical biochemistry laboratories. While the book presents the basic clinical biochemistry laboratory techniques in easily understandable language, it also introduces some higher-level concepts to prepare the students for the future.

The help and assistance provided by Dr. Jagriti Narang, assistant professor at Jamia Hamdard University, New Delhi, at each and every step during the writing of this book is enormous. We thank her and her family members for their support. We also thank Jenny Stanford Publishing and its team for accepting the book and assisting in the publication process.

We hope the book will meet the readers' requirements and will be glad to accept constructive criticisms and suggestions from the faculty, students, and readers to make this book a better one in the future.

Rooma Devi
Aman Chauhan
Simmi Kharb
Chandra Shekhar Pundir
Summer 2023

Chapter 1

Lab Safety Rules

Cardinal Rules to be Followed in the Laboratory

- Tell the lab in charge if any accident happens, including spills.
- Tie back long hair before beginning a lab investigation.
- Always wear gloves, safety goggles, and aprons when working with chemicals, heat, and flames.
- Do not eat or drink anything during a lab investigation.
- Use tools only as they are supposed to be used. Treat tools and materials with respect. Make sure they are clean before and after you use them.
- Always clean your work area after an investigation.

Good Laboratory Techniques

- Students should read the relevant data regarding the experiments to be performed.
- Do not operate new unfamiliar equipment until you have been given instructions about them.

Clinical Biochemistry: A Laboratory Guide
Rooma Devi, Aman Chauhan, Simmi Kharb, and Chandra Shekhar Pundir

ISBN 978-981-4968-75-1 (Hardcover), 978-1-003-45566-0 (eBook)
www.jennystanford.com

- All the instructions and labels in the laboratory should be carefully read.
- Strong acids and alkalis should never be mouth pipetted. Make sure the acid and alkali should not fall on your hands or any other body parts.
- Autopipettes and dispensers should be used.
- Note down your observations of qualitative experiments as test, observation, and inference and quantitative experiments as aim, principle, observation, calculation, reference range, and clinical significance.

Laboratory Hazards

Biological Hazards

- Every patient specimen should be treated as potentially infectious.
- Blood samples from high-risk patients, such as hepatitis B and AIDS, should be collected, transported, handled, and processed under strict precautions.
- Specimens should remain "capped" during centrifugation to prevent the formation of infective aerosols.

Physical and Chemical Hazards

- Careless handling of apparatus and reagents is the common cause of laboratory accidents, resulting in burns or fires, which must be reported and treated promptly as per the chart displayed in the laboratory.
- In a clinical biochemistry laboratory, all essential elements for a fire to begin are present, such as an ignition source; hence, it is important to assure the safety of self, personnel, and equipment.

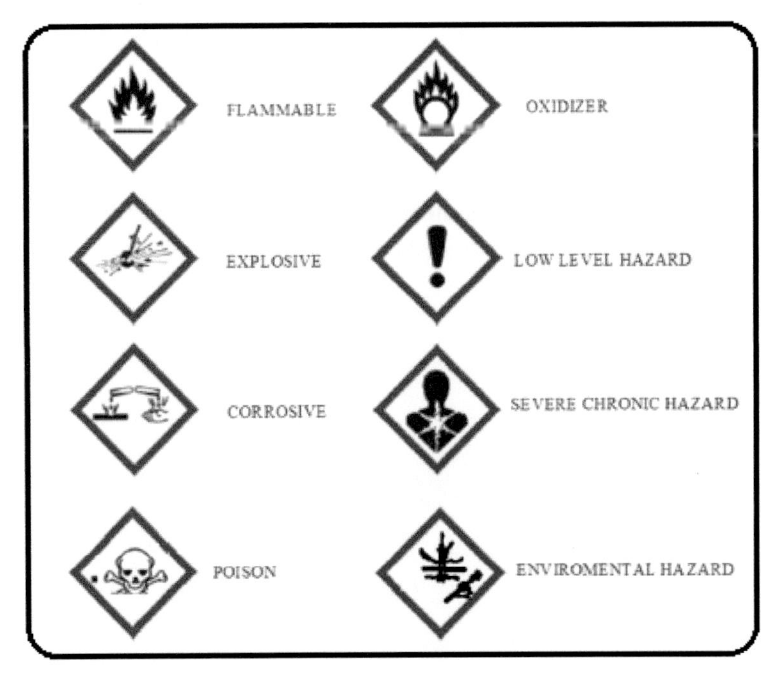

Figure 1.1 Signs and symbols used in a laboratory.

Fire Extinguishers

They are used to handle fire and possess dry or wet chemicals.

Fire extinguishers are classified into four types:

1. Class A extinguishers are used to handle fire which has happened due to the rubber, wood, clothes, etc.
2. Class B extinguishers are used to handle fire whose root cause is flammable liquids such as alcohol, etc.
3. Class C extinguishers are used to handle fire caused by electrical sources.
4. Class D extinguishers are used to handle fire whose root cause are flammable metallic substances such as sodium, etc.

Laboratory Glassware

Beakers

Beakers are used mainly for the preparation of reagents and solutions. They are straight-sided cylindrical vessels available in a wide range of volumes from 5 ml to several liters.

Flasks

Flasks have capacities of 25–500 ml. Different types of flasks are available:

- **Conical flasks (Erlenmeyer type):** These flasks are used for performing titrations and boiling solutions. Evaporation is minimum in these flasks because of the conical shape.

- **Flat-bottomed round flasks:** These flasks are used mainly for heating liquids.

- **Round-bottomed flasks:** These flasks can withstand high temperatures. So, they are used for evaporating samples to dryness and for the distillation of water, alcohol, and other organic compounds.

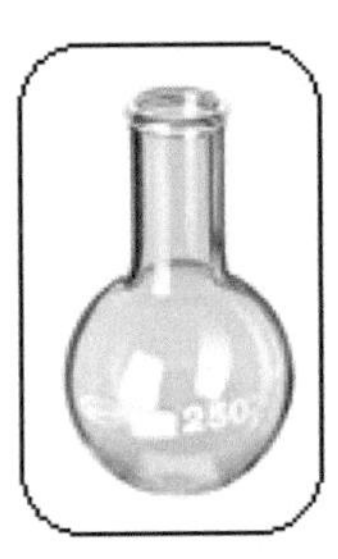

- **Volumetric flasks:** They are flat-bottomed, pear-shaped vessels having long narrow necks with a specific volume mark and fitted with a stopper.

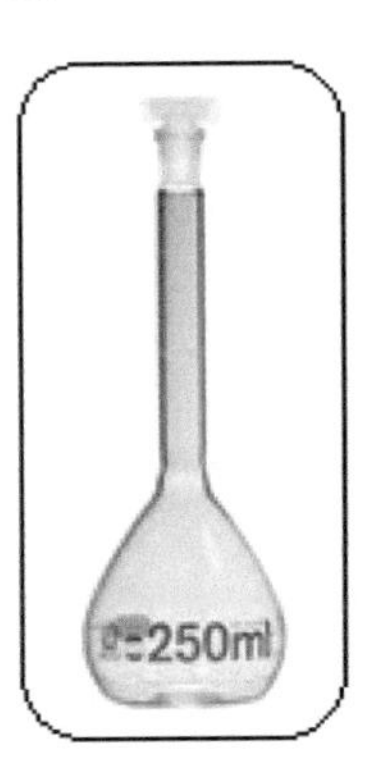

Graduated (Measuring) Cylinders

Graduated (measuring) cylinders are narrow, straight-sided vessels that are used to measure specific volumes. They are available in sizes ranging from 10 ml to several liters. A high degree of accuracy is not possible because of their wider bore.

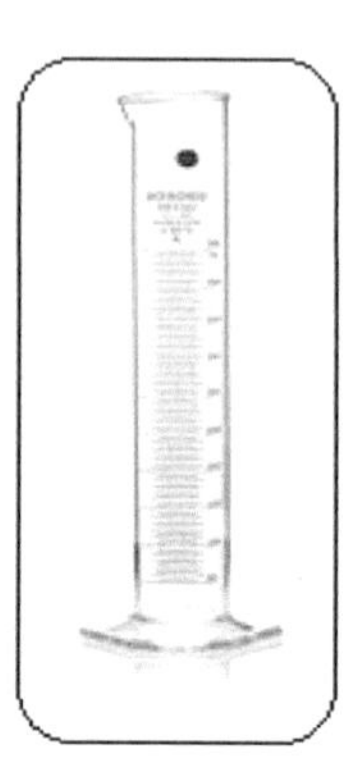

Burettes

Burettes are long, graduated tubes with a stop cork at one end, available in capacities of 10 to 100 ml. These devices are used to accurately deliver known volumes of liquids into a container. By measuring from one graduated line to another graduated line, one can deliver even fractional volumes (less than 1 ml) of liquids with a high degree of accuracy. They are used mainly for titrations and also for dispensing corrosive reagents.

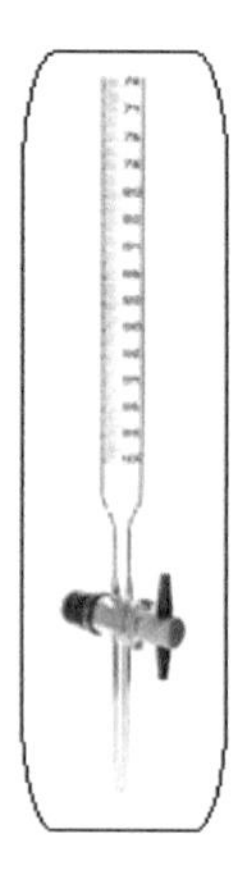

Funnels

Funnels are used for transferring liquids/solids into containers and for separating solids from liquids. They usually have short or long, thin stems. They are used with filter paper to remove particles from solutions. Funnels with wide-mouthed stems, which allow solids to pass through easily, are used for transferring solids into containers.

Bottles

Reagent Bottles

These bottles are cylindrical with a narrow neck fitted with a stopper, available in various capacities of 25–2000 ml. They are made up of plain, white, or amber-colored glass. Amber-colored bottles are useful to store certain light-sensitive chemicals such as silver nitrate.

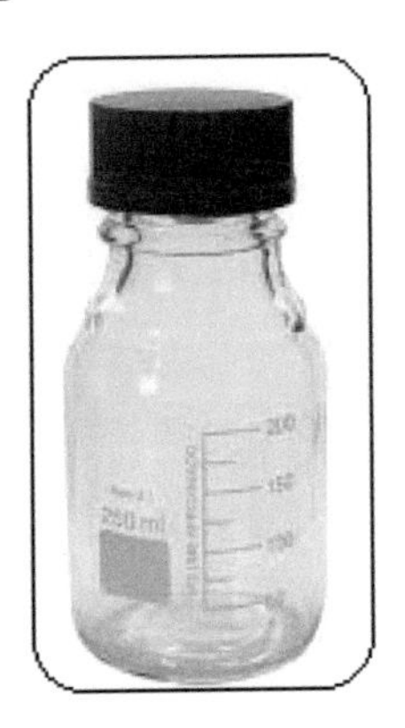

Drop Bottles

These bottles have a narrow neck with a slotted glass stopper, available in 50–100 ml capacities. They are used for the delivery of drops of solutions such as stains and indicator solutions and are made up of white or brown glass.

Wash Bottles

These bottles are usually plastic bottles with a delivery tube at the top.

Pipettes

Pipettes are used for the delivery of accurate and controlled volumetric measurements. They are of various types differing in their levels of accuracy and precision, which includes complex adjustable or automatic pipettes.

Manual Pipettes

- **To deliver (TD) type of pipettes:** These pipettes must be held vertically, and the tip must be placed against the side of the accepting vessel to drain liquid by gravity. Common pipettes included under the TD type are graduated and volumetric pipettes.

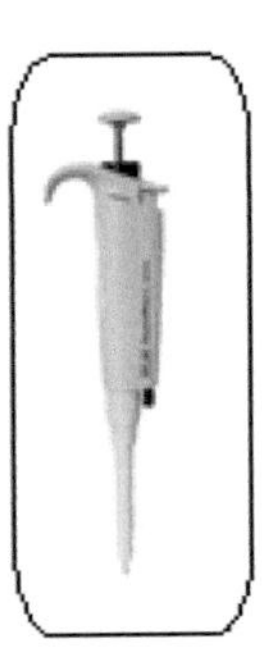

- **Graduated pipettes:** These pipettes are available from 0.1–10 ml capacity, e.g., Mohr pipettes and serological pipettes. Mohr pipettes are glass tubes of uniform diameter with a tapered delivery tip and have graduations made at uniform intervals but well above the tapered delivery tip. They are mainly used for pipetting distilled water and reagents. However, 0.1 ml and 0.2 ml pipettes are used for pipetting specimens such as blood, serum, or plasma. Serological pipettes are of either TD or blowout types.

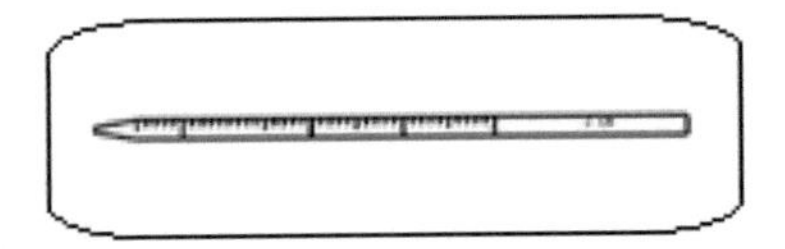

- **Volumetric pipettes:** These pipettes are not graduated but designed specially to deliver a specific quantity of a specimen. They have an open-ended bulb holding the bulk of the liquid; a long glass tube at one end, which has the mark to describe the extent to which the pipette is to be filled; and a tapered delivery portion. These pipettes hold and deliver only the specific volumes indicated at the upper end of the pipette. They are used mainly to pipette specimens and standards and are very accurate. Pipettes must be refilled or rinsed out with the appropriate solvent after the initial liquid has been drained from the pipettes.

- **Micropipettes:** They can deliver volumes ranging from 1 to 500 µl. A micropipette consists of capillary tubing with a line demarcating a specific volume. It is filled to the line by capillary action.

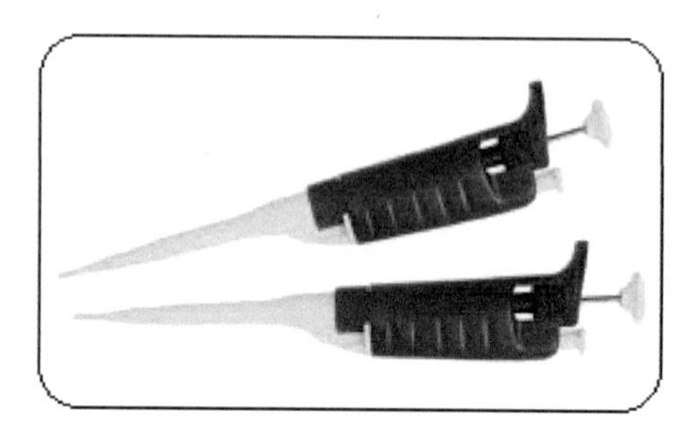

- **Pasteur pipettes:** These pipettes have a rubber bulb attached to the top of a glass tubing. They are tapered at the tip and are especially useful in delivering urine samples.

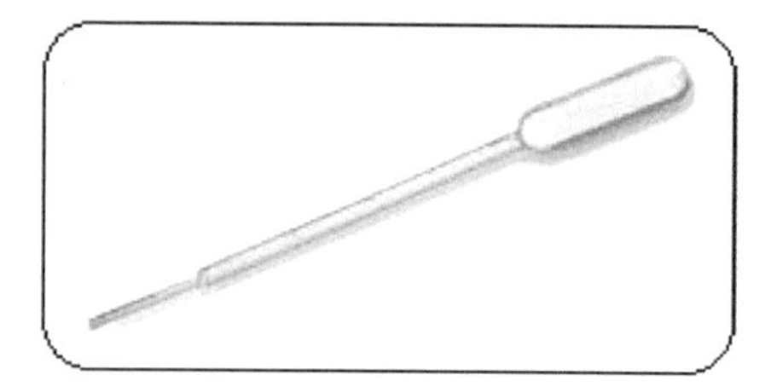

Auto Pipettes

Sucking and blowing with the mouth are not done in auto pipettes. A mechanical plunger does this work. They are frequently used in the laboratory to repeatedly add a specific volume of a reagent. They are mainly of push-button type (Eppendorf type) and are piston-operated devices to dispense liquids. These pipettes can be of fixed volume type or variable volume adjustable type.

- **Fixed volume type:** The volume of the fluid sucked is fixed. Different pipettes are used to pipette different fixed volumes. They can dispense fixed volumes of 10 µl, 20 µl, 50 µl, 100 µl, 200 µl, and 1000 µl as required.
- **Variable volume adjustable type:** The volume of fluid to be dispensed can be adjusted with the adjusting screw as

required. Variable volumes, such as 20–200 μl and 100–1000 μl, are available.

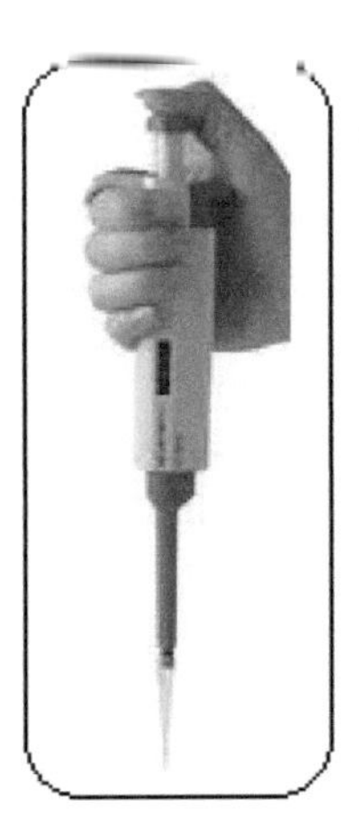

Test Tubes

Test tubes are of uniform thickness, which can withstand mechanical and thermal shocks. Tubes with a rim are preferred when a reagent in a test tube is directly heated on the flame using a test tube holder. Test tubes are available in capacities of various volumes.

Outer diameter × length (mm):

- 10 × 75 mm: Used for testing, identification of biochemical substances, and centrifugation.
- 15 × 125 mm: Used for most biochemistry tests.
- 18 × 150 mm: Used for heating a reaction mixture directly on a flame.

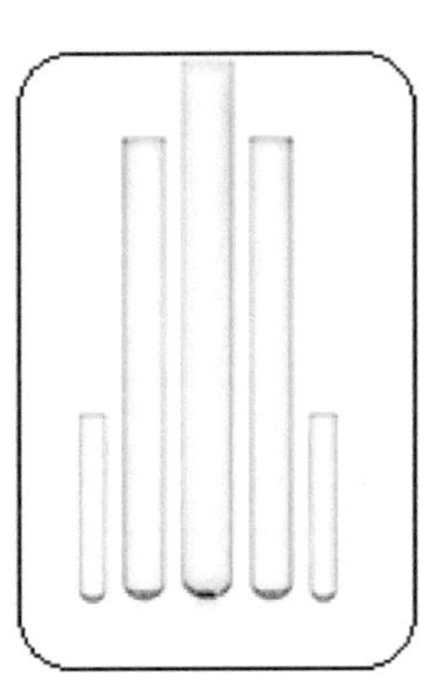

Centrifuge Tubes

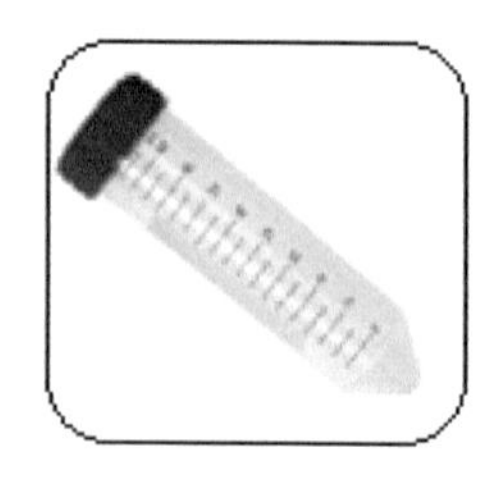

Centrifuge tubes are either graduated or plain. They are usually conical in shape, and their size is usually 17 × 120 mm.

Folin-Wu Tubes

Folin-Wu tubes have markings at 12.5 ml and 25 ml; they have a bulb at the bottom with a constriction. They are used for the determination of blood sugar by the Folin-Wu method.

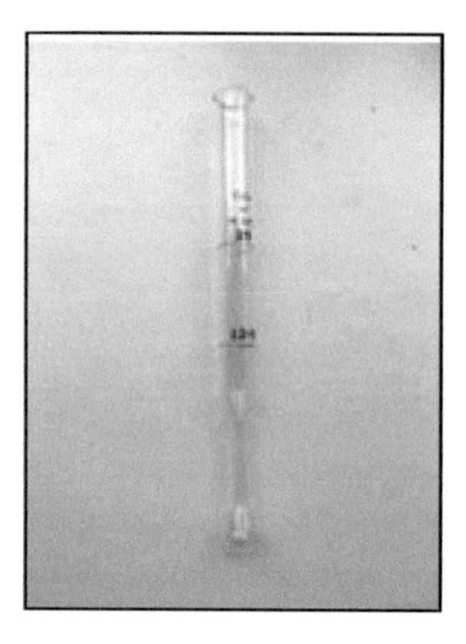

Dispensers

Dispensers are used to dispense large, fixed volumes of reagents. They are usually used to dispense strong acids and alkalis.

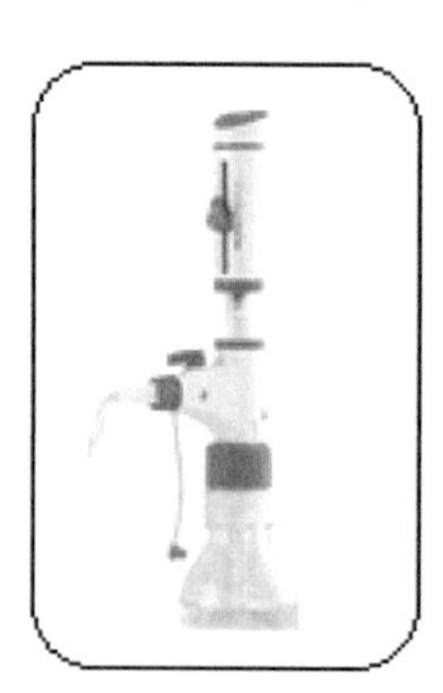

Desiccators

Desiccators are used to keep solid or liquid materials dry. They usually have an area at the bottom where a desiccant (water-absorbing material) is placed, which removes the water of hydration from compounds.

Cleaning of Glassware

Glassware should be thoroughly rinsed with tap water and cleaned with some detergents. Finally, they should be rinsed with tap water followed by distilled water. An apparatus can be dried quickly by rinsing with alcohol followed by ether. The cleaned glassware, except graduated glassware, is dried in a hot air oven. A dichromate–sulfuric acid mixture (chromic acid) is used for cleaning glassware, which removes even the last traces of grease.

Disposal of Laboratory Waste

- **Biomedical waste:** Any waste generated during the diagnosis, treatment, or immunization of human beings or animals or in research activities. Workers in clinical laboratories are exposed to biological hazards due to the handling of infected specimens. Proper safety is required in not only handling but also disposing of these infected materials. Waste should be segregated at the point of generation and disposed of in bags with the correct color.

- **Yellow bins/bags:** Human anatomical wastes such as tissues, body parts, swabs, and items contaminated with blood and body fluids, discarded gloves, etc.

- **Red bins/bags:** All infected plastic recyclable wastes such as waste generated by tubings, urine bags, etc.

- **Black bins/bags:** Noninfectious general wastes such as paper, peels, and wrappers.

- **Blue bins/bags:** All glass wastes such as needles, syringes, and sharps are first destroyed in a needle destroyer and discarded in a sharp-disposal unit containing 1% bleach.

Practical Spots in Biochemistry

Spot 1

1. Name the glassware shown in the picture above?
2. Mention its uses.
3. Name the different types of glassware used in a biochemistry laboratory.
4. What material is used for making glassware for laboratory use?

Spot 2

1. Identify the picture shown above.
2. What are the different types of flasks?
3. What is the material used to make this flask?
4. What is the use of the glassware shown above?

Spot 3

1. What is shown in the above picture?
2. What is biomedical waste?
3. Why there is a need for biomedical waste management?
4. Bags of what color code are used for disposing of glassware?

Chapter 2

Specimen Collection and Processing

Proper collection, processing, storage, and transport of samples are required to get the best quality results. Proper identification of patients is required for proper sampling, which can be done by matching the names, medical record numbers, birth dates, and photographs (in medicolegal cases).

The process of collecting blood is called phlebotomy, and the person who performs phlebotomy is called a phlebotomist. Biological samples such as blood, saliva, urine, and feces are analyzed for various parameters, which help clinicians to diagnose and treat various diseases. Improper sample collection and handling can cause hemolysis, which involves the disruption of the red blood cell membrane and the release of hemoglobin. Hemolysis may be in vivo (inside the body) or in vitro (outside the body).

Specimens Analyzed in Clinical Laboratories

Specimens that are analyzed in clinical laboratories are as follows:

- Whole blood
- Serum
- Plasma
- Urine
- Feces

Clinical Biochemistry: A Laboratory Guide
Rooma Devi, Aman Chauhan, Simmi Kharb, and Chandra Shekhar Pundir

ISBN 978-981-4968-75-1 (Hardcover), 978-1-003-45566-0 (eBook)
www.jennystanford.com

- Saliva
- Other fluids such as pericardial, amniotic, synovial, etc.

Blood for analysis may be taken from the following sites:

Venipuncture

- Antecubital fossa (preferred)
- Back of hand
- Ankle

Skin Puncture

- Tip of the finger
- Heel of infants (medial and planter surface of foot)

Arteril Puncture

- Radial artery
- Brachial artery in the elbow
- Femoral artery in the groin

Skin puncture is used in special conditions such as severe vein damage due to the repeated venipuncture/limited volume of blood, such as in pediatric patients/burns or in case a sample is to be applied directly to a testing device.

Arterial punctures need a skilled hand and must be done by a trained personal. These punctures are mainly used for blood gas analysis.

Venipuncture

The puncture of a vein to withdraw blood for analysis or intravenous injection is called venipuncture. A phlebotomist should wear proper PPE (gloves, gown). The extent of precaution varies with the kind of patient illness. Ask about allergy from latex, etc. The position of patients should be supine or seated. Infants and young children should be held to prevent movements, which may cause injury. A needle of appropriate size should be used (adults—19 to 23 gauge and pediatric patients—23 to 25 gauge). Venipuncture should be done from the antecubital fossa (preferred site) or from the back of hand or ankle. In infants and children, dorsal hand veins are preferred. The area around the puncture site should be cleaned with 70% isopropanol/benzalkonium chloride. Venous occlusion should be attained by tourniquet/blood pressure cuff (tourniquet should be applied 4 to 6 inches above the intended puncture site). Blood should be collected in the desired vacutainer, and the order of draw must be followed. A dry gauge pad should be used over the puncture site to stop bleeding and promote the clotting process.

Vacutainers

Vacutainers are sterile glass or plastic tubes having colored rubber stoppers, which create vacuum inside the tube. The size of vacutainers may vary for adults and pediatric patients. They contain different types of anticoagulants and preservatives. Blood starts clotting a few minutes after it is removed from the body. The normal process of clotting may be stopped by the addition of anticoagulants. Serum is the fluid portion of the clotted blood after centrifugation, whereas plasma is the fluid portion of the unclotted blood.

Red tube:

- Serum separating tube
- No additive
- For example, antibodies and drugs

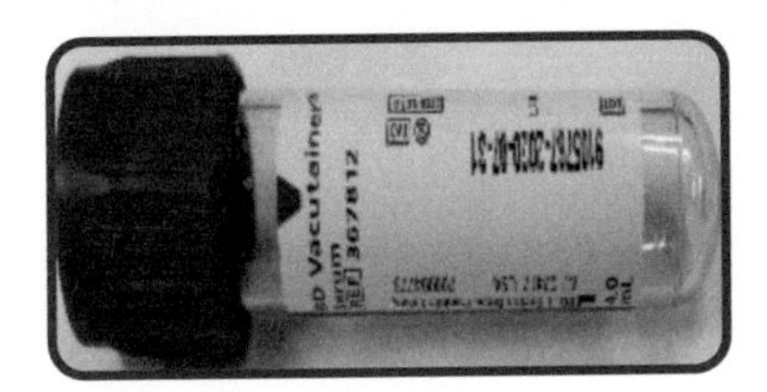

Light blue tube:

- Sodium citrate
- Used for clotting (coagulation) studies
- Plasma separating tube
- Must be completely filled

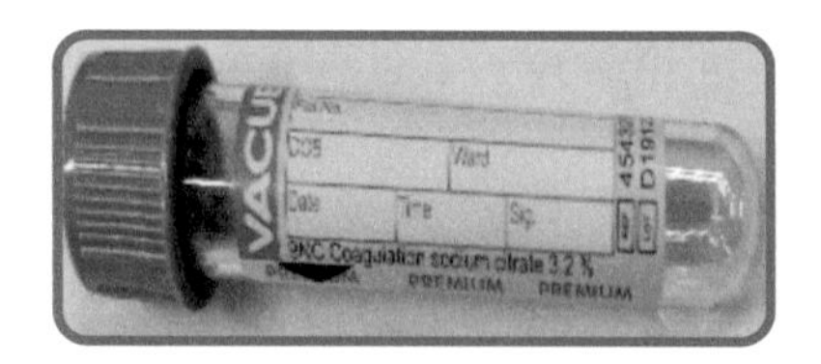

Yellow tube:

- Whole blood (reference testing)
- Acid citrate dextrose as anticoagulant
- For example, blood culture

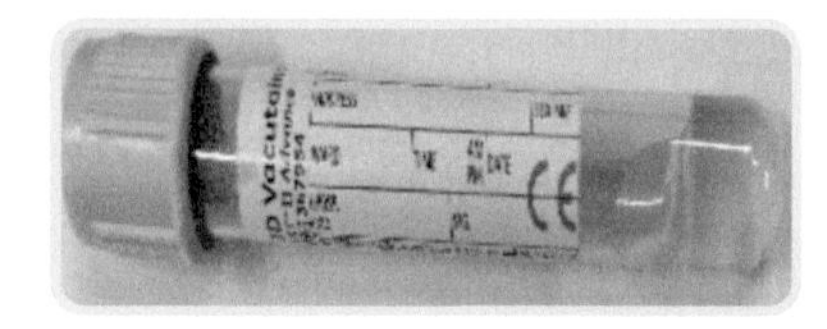

Gray tube:

- Potassium oxalate + sodium flouride
- Used for measuring glucose level

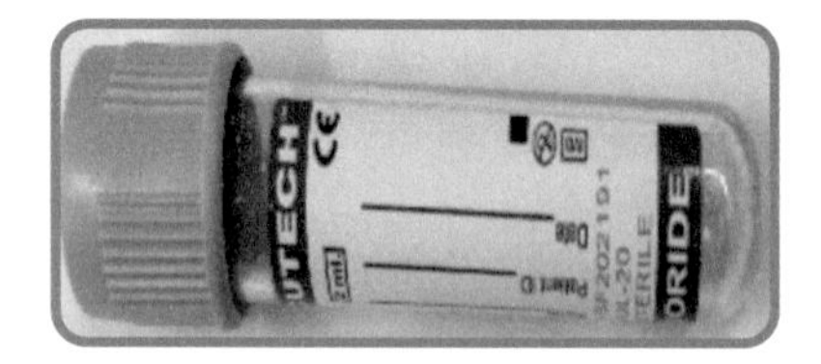

Lavender tube:

- Strong anticoagulant EDTA (ethylene diamine tetra acetic acid)
- Used for hematology studies

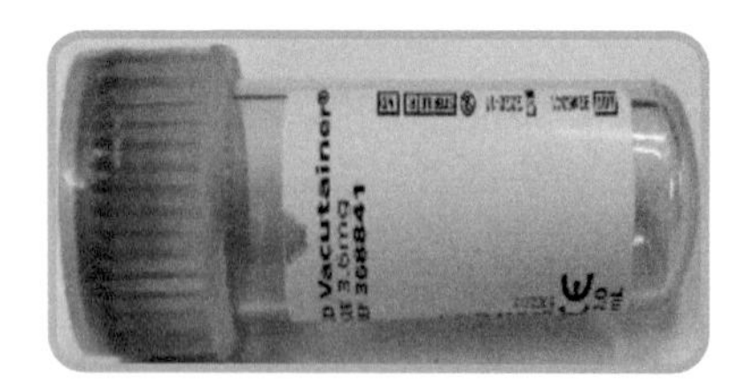

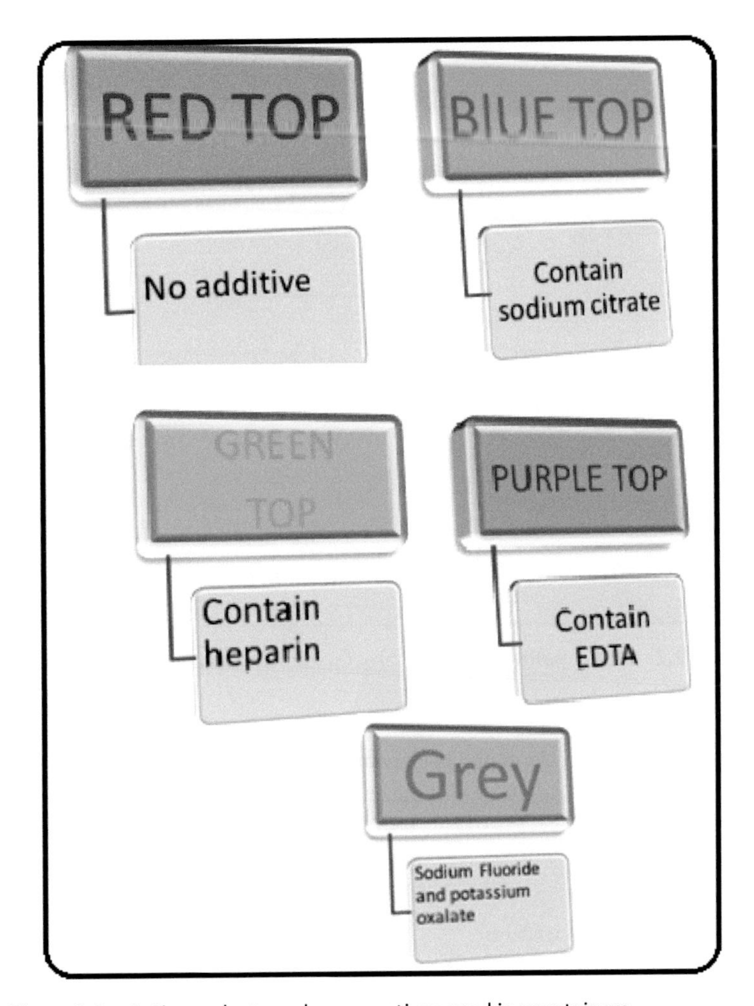

Figure 2.1 Anticoagulants and preservatives used in vacutainers.

Anticoagulants and Preservatives Used in Vacutainers

For a better understanding of anticoagulants, proper knowledge of the blood coagulation pathway is essential. Any breach in the steps of the coagulation pathway leads to non-clotting of blood.

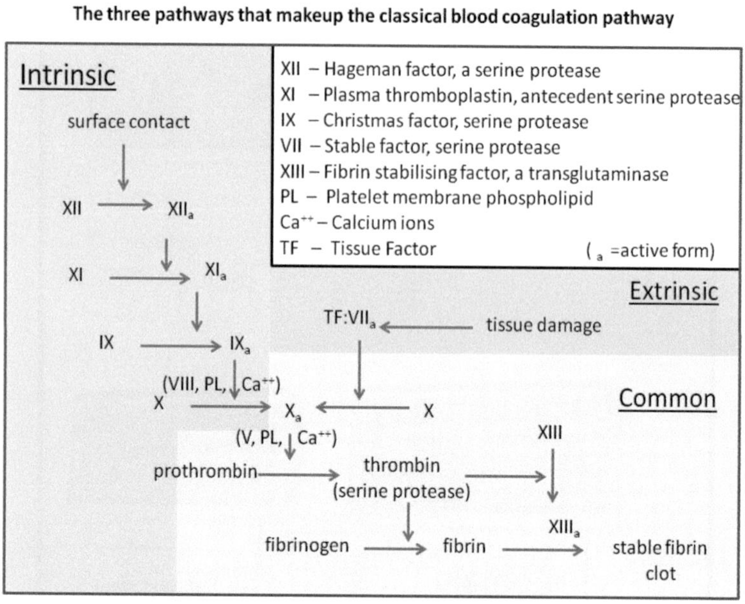

Figure 2.2 Three pathways that make up the classical blood coagulation pathway.

Table 2.1 Anticoagulants and preservatives with their actions

Anticoagulant/Preservative	Action
Heparin	Prevents the formation of fibrin from fibrinogen
Ethylenediaminetetraacetic acid (EDTA)	Prevents coagulation by binding calcium
Sodium fluoride (added as preservatives for blood glucose and lactate)	Inhibits glycolysis
Citrate	Chelates calcium
Oxalate	Forms an insoluble complex with calcium ions
Iodoacetate	Antiglycolytic agent (substitute for NaF)

Urine Analysis

The most important factor in a urine sample is the timing of collection.

- Abnormal constituents: Clean, early morning, fasting specimen (most concentrated specimen)
- Bacterial examinations: First 10 ml of urine
- Bladder disorders: Midstream specimen
- Critically ill patients: Catheter/suprapubic specimen, etc.
- Infants: Bagging

In a long-standing urine specimen, there might be some changes such as bacterial action, urinary decomposition, and precipitation of phosphates. Preservation of urine is required for good-quality results. Preservatives used depend on the parameters to be analyzed in urine. Refrigeration is the best acceptable method. Acidification of urine to preserve specimen for 24 hours may be done, which is further used to determine calcium, steroids, adrenaline, etc. A mild base is used to preserve porphyrins and uric acid.

Practical Spots in Biochemistry

Spot 1

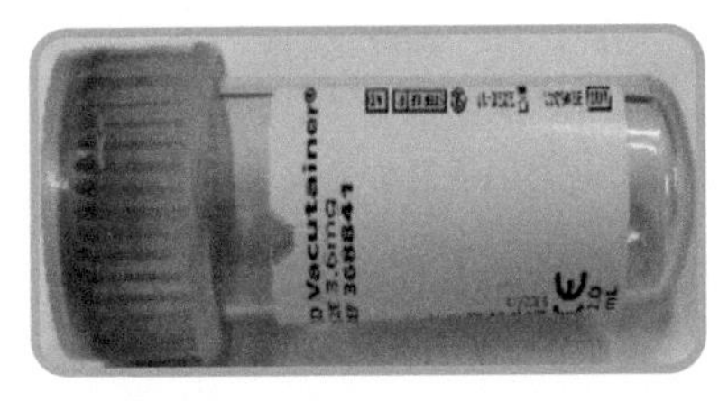

1. Name the device in the picture shown above.
2. Mention its uses.
3. List the additive present in this tube.
4. What is the mechanism of action of the additive?

Chapter 3

Quality Management in the Laboratory

Quality Control

Quality control (QC) programs were usually concerned with making sure that the results furnished by test tactics are as correct as possible. Controls are commercially available assayed solutions that include constituents identical to those being analyzed in the patient sample. A normal control product contains a normal range of analytes, whereas an abnormal control contains below or more range of analytes. Generally, in a clinical laboratory, multiple controls may be used. But in practice, it has been seen that two to three controls are mostly used. It is better to use two controls, one having the normal reference interval and the other one outside the reference interval.

Internal Quality Control

Internal quality control is performed on a regular basis to assess the performance of measuring systems.

External Quality Control/Proficiency Testing

Performance is assessed by comparing the obtained results with those from different laboratories. When the QC results are within

Clinical Biochemistry: A Laboratory Guide
Rooma Devi, Aman Chauhan, Simmi Kharb, and Chandra Shekhar Pundir

ISBN 978-981-4968-75-1 (Hardcover), 978-1-003-45566-0 (eBook)
www.jennystanford.com

the acceptable range, only then the samples of the patients to be analyzed should run. The results of the patients must not be released unless QC results are below the acceptable range.

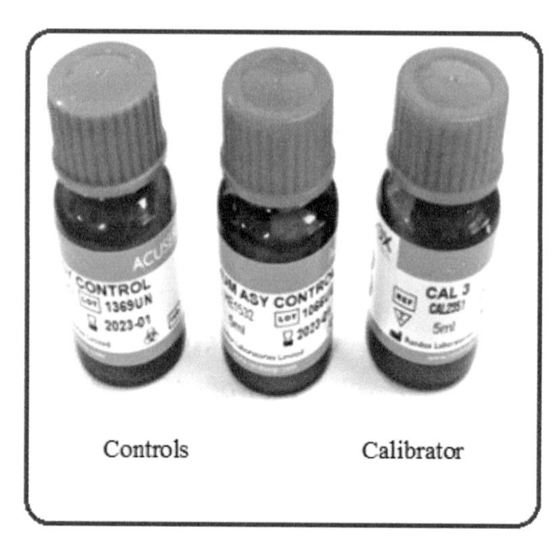

Figure 3.1 Controls and calibrators used in the laboratory.

The common terms used in a clinical laboratory are as follows:

Standard

- It is a substance having the exact known concentration.
- It is used to calibrate a new instrument or to recalibrate after repairing an old one.
- It is used if a problem appears with any test method.
- It is also called the reference material.

Calibration

- Calibration is the process of standardizing or adapting a method or tool for accurate results.

Random Errors

- For QC results, positive or negative deviations from the calculated mean are defined as random errors.
- These errors may be acceptable or not acceptable depending on the standard deviations (1SD, 2SD, 3SD, etc.).

Specificity

- Specificity is the ability of an analytical method to determine only the analyte to be measured.

Sensitivity

- Sensitivity is the ability of analytical methods to detect trace amounts of the analyzed material being measured.

Systemic Errors

- Systemic errors are responsible for consistent higher or lower results than normal.
- They may result due to instrument faults/expired reagents.

Accuracy and Precision

Accuracy is the closeness of the estimated value to that of the true value. Precision is the closeness of the results on repeated analysis.

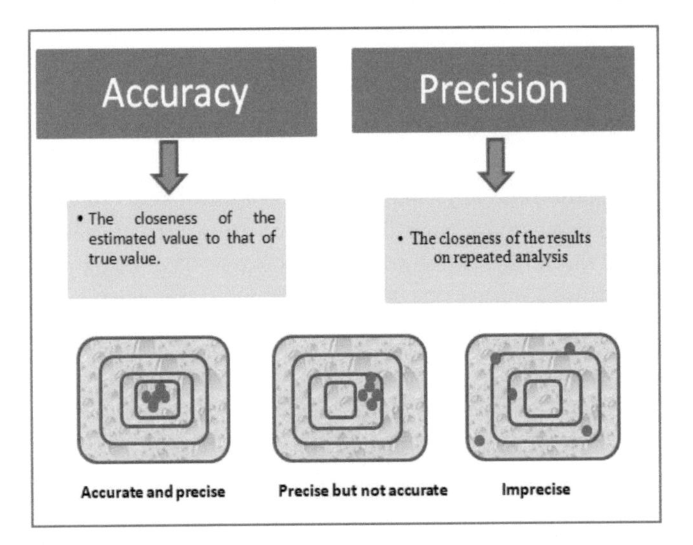

Mean

Mean is the arithmetic average of the results of repeated measurements of the quality control sample.

$$\text{Mean} = \Sigma X_n / n,$$

where

Σ = Sum

X_n = Each value in the data set

n = Total number of values

To calculate the mean, first add all values of the control divided by the total number of values.

Standard Deviation

Standard deviation is a measure of the amount of variation or spread of a set of values. It is calculated by the following formula:

$$\text{SD} = \sqrt{\frac{\sum (x_n - \bar{x})^2}{n-1}}$$

where $\sum (x_n - \bar{x})^2$ is the sum of the squares of differences between individual QC values and the mean and n is the total number of values.

Levey–Jennings Chart

The Levey–Jennings (L–J) chart gives a graphical representation of the control values with time. The control values are plotted on the x-axis, and the dates are plotted on the y-axis. The y-axis provides the target values with ±1SD, 2SD, and 3SD. So, standard deviations help in plotting the L–J chart.

Table 3.1 West guard rules used in a clinical biochemistry laboratory

	Quality Control Evaluation Rules
1_{2s}	Control observations exceed ±2s on the mean.
1_{3s}	Control observations exceed ±3s on the mean. High sensitivity is allowed to random error.
2_{2s}	Two control observations across the same side exceed, i.e., +2s or −2s. High sensitivity to system errors is allowed.

R_{4s} One control exceeds +2s, and another exceeds −2s. The detection of random errors is allowed.

4_{1s} Four consecutive control observations exceed +1s or −1s. This allows the detection of systemic errors.

10_x Ten consecutive control observations fall on one side or the other of the mean (no requirement for SD size). This makes it possible to recognize systemic errors.

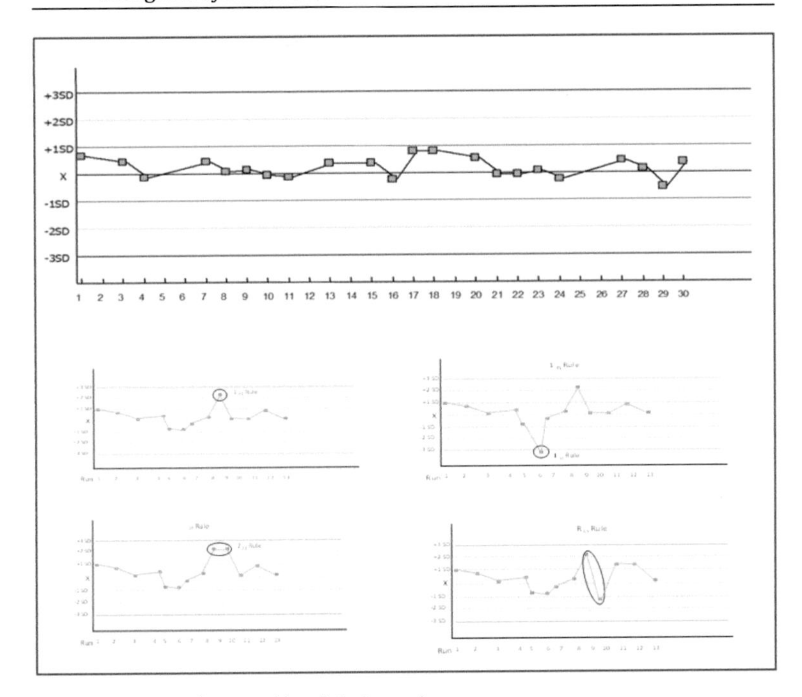

Figure 3.2 L–J chart and its violation rules.

Coefficient of Variation

The coefficient of variation (CV) is also called the relative standard deviation. It is standard deviation expressed as the mean percentage.

$$CV = (SD/X)\ 100$$

Questions and Answers

1. What is the difference between a calibrator and a standard?

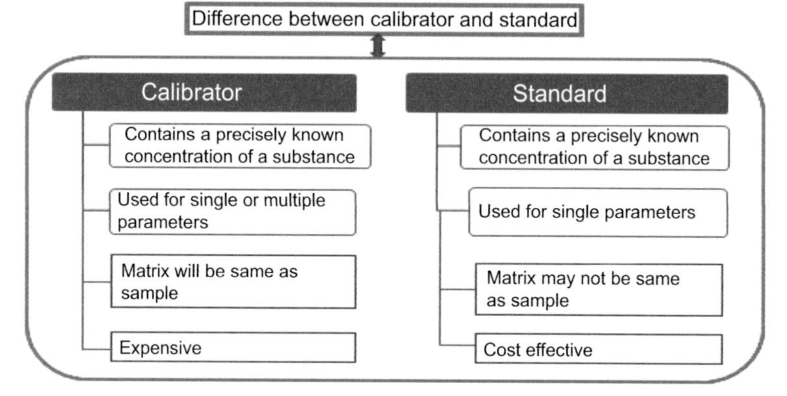

2. What are end point assay, kinetic assay, and fixed time kinetics?

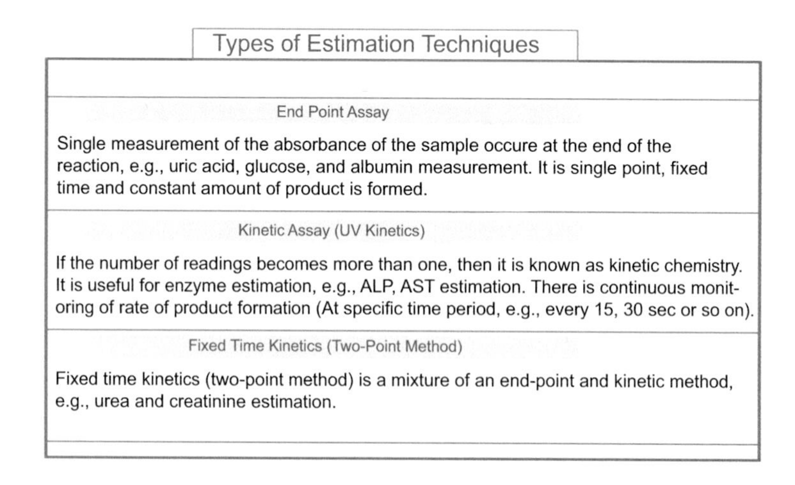

Types of Estimation Techniques
End Point Assay Single measurement of the absorbance of the sample occure at the end of the reaction, e.g., uric acid, glucose, and albumin measurement. It is single point, fixed time and constant amount of product is formed.
Kinetic Assay (UV Kinetics) If the number of readings becomes more than one, then it is known as kinetic chemistry. It is useful for enzyme estimation, e.g., ALP, AST estimation. There is continuous monitoring of rate of product formation (At specific time period, e.g., every 15, 30 sec or so on).
Fixed Time Kinetics (Two-Point Method) Fixed time kinetics (two-point method) is a mixture of an end-point and kinetic method, e.g., urea and creatinine estimation.

Chapter 4

Urine Analysis

Urine

Urine is formed by the kidneys. It is used in biochemistry laboratories for biochemical analysis. Many dangerous waste materials and water are eliminated from the body via urine. It is a regular physiological system of the body.

Urine analysis provides mainly diagnostic information on conditions such as

- Metabolic disorders
- Liver disorders
- Kidney disorders
- Endocrine disorders
- Pregnancy
- Urinary tract infections, etc.

A complete urine analysis can be done by the following:

- Biochemical analysis
- Bacteriological analysis
- Microscopic analysis

As in biochemistry, complete biochemical analysis is to be done, we will discuss about only the biochemical analysis of urine.

Clinical Biochemistry: A Laboratory Guide
Rooma Devi, Aman Chauhan, Simmi Kharb, and Chandra Shekhar Pundir

ISBN 978-981-4968-75-1 (Hardcover), 978-1-003-45566-0 (eBook)
www.jennystanford.com

Biochemical Analysis

Biochemical analysis involves

- Qualitive estimation of abnormal constituents
- Quantitative estimation of abnormal constituents

Urine Sample Types

- First morning urine samples (for pregnancy test)
- Second morning urine samples (for glycosuria, microalbuminuria)
- Random urine samples (for chemical urine test and sediment, amylase, etc.)
- Twenty-four-hour urine samples (for wastes in urine such as Na, K, proteins, hormones, etc.)
- Short-term urine samples (4 h) (for renal function test)

Table 4.1 Routine physical examination of urine

Experiment	Observation	Inference
Appearance (color, transparency)	Clear	The given sample of urine is normal.
Odor	Aromatic smell	The given sample of urine is normal.
Specific gravity	1.016 to 1.025	Normal
Volume	750/2000 ml/day	Normal
Reaction to litmus	Blue litmus turns red	Normal urine is acidic.

Color: The color of urine is usually described after visual inspection with common color terms.

- The normal color of urine is straw due to urochrome.
- The normal color of urine changes due to different disease conditions or in persons taking medications.
- **Light yellow to colorless urine:** Diabetes mellitus, diabetes insipidus (cause polyuria)
- **Dark yellow:** Jaundice, use of vitamin B complex
- **Amber:** Dehydration during burns or fever

- **Yellow orange:** Urobilin, bilirubin
- **Yellow brown/yellow green:** Biliverdin
- **Red/brown:** Hematuria, hemoglobinuria, porphyria
- **Brown/black:** Alkaptonuria, melanoma
- **Green/blue:** Bacterial infections, drugs (methylene blue, etc.)

Table 4.2 Urine analysis based on color

Color	Metabolite	Clinical Condition
Red	RBC, hemoglobin	Hematuria Hemoglobinuria
Red brown	Myoglobin Porphyrin	Myoglobinuria Porphyrinuria Menstrual contamination
Yellow	Urochrome	Healthy
Yellow orange	Urobilin Bilirubin	Dehydration Jaundice
Yellow green	Biliverdin	Jaundice
Brown black	Homogentisic Methemoglobin	Alkaptonuria
Milky white	Chyle	Chyluria

Transparency:

- Freshly voided urine is transparent and clear.
- The presence of proteins, pus cells, blood, bacteria, urates, and phosphates makes the urine turbid.

Specific gravity: Concentrated urine has a high specific gravity, while diluted urine has a low specific gravity. It is measured by a urinometer or refractometer.

Urine has high specific gravity due to the following reasons:

- Hepatic diseases
- Diabetes mellitus
- Congestive heart failure
- Excessive lack of water because of sweating, fever, vomiting, and diarrhea
- Nephrosis

Urine has low specific gravity due to the following reasons:

- Diabetes insipidus
- High fluid intake
- Glomerulonephritis

Urine shows fixed specific gravity due to the following reasons:

- Several renal damages with disturbance in both the concentrating and diluting abilities of the kidney.

Volume

Polyuria: A urine output of more than 3 L/day is termed polyuria. It is the increased excretion of urine. Polyuria occurs in the following conditions:

- Diabetes insipidus
- Diabetes mellitus
- Chronic renal damage

Oliguria: A urine output of about 500 ml/day is termed oliguria. It occurs in the following conditions:

- Fever
- Cardiac failure
- Shock
- Less fluid intake
- Excessive fluid loses due to vomiting, diarrhea, and sweating

Anuria: Anuria is the suppression of urine output (generally less than 50 ml per 12 hours). It is the total loss of urine.

Reaction (pH): Normal fresh urine is acidic in nature, with pH 6 approximately. The pH of urine may range from 4.5 to 8. A pH paper is commonly used for the measurement of pH.

Acidic urine: A pH of 4.5 or less may be seen due to the following reasons:

- Chronic respiratory distress
- High protein diet
- Metabolic acidosis
- Fever

Alkaline urine: A pH of 8.5 or more may be seen due to the following reasons:

- Chronic renal failure
- Urinary tract obstruction
- Renal tubular acidosis
- Heavy meals

Odor

- Freshly voided urine is slightly aromatic.
- Urine becomes more ammoniacal due to bacterial activity.
- A fruity odor is felt in severe diabetes due to the presence of ketone bodies.
- A putrid or foul smell is felt during urinary tract infections.

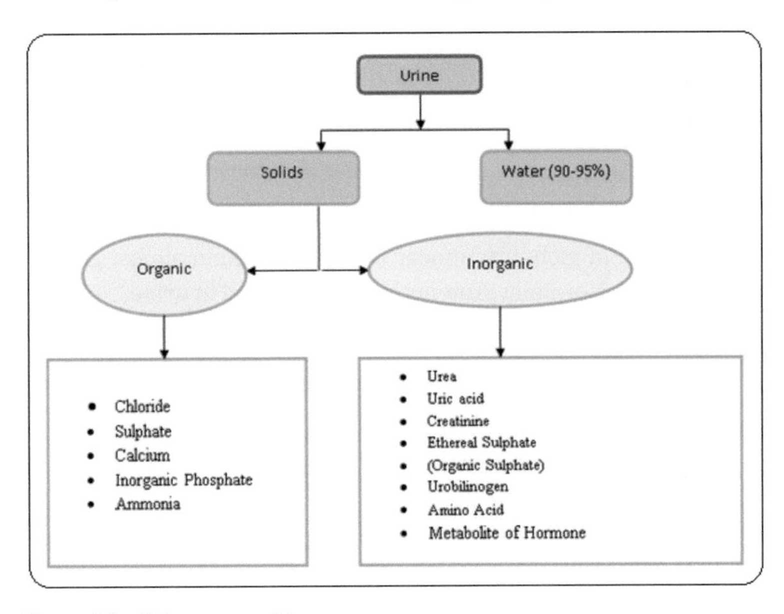

Figure 4.1 Urine composition.

Chemical Examination of Urine

Chemical analysis of urine is conducted to detect the presence or absence of certain constituents that reflect the disease state of a body. The parameters include proteins, sugar, blood, ketone bodies, bile pigments, bile salts, and urobilinogen. Conventional test methods are commonly employed. Reagent test strips (dip sticks) are commercially available and recently used.

Proteins

- Normally a minute amount of protein (100 mg/day) is excreted and is not detectable by routine qualitative test methods.
- Proteinuria is the term used for increased protein excretion in urine, and it is seen in kidney disease, such as nephrotic syndrome, multiple myeloma, chemical poisons, etc.
- A commonly found protein in urine is albumin, so it is often termed albuminuria.

Methods of Protein Detection

- Heat coagulation test
- Sulfosalicylic acid test
- Nitric acid test
- Reagent test strips (dip sticks)

Albumin and globulins appear in urine. Albumin almost greatly predominates. Normally proteins are not excreted in urine.

Causes of Albuminuria

- **Pre-renal causes:** No primary kidney disease
  - Drugs and chemical poisoning
  - Intra-abdominal tumors
  - Fever and toxemia
- **Renal causes:** Primary kidney disease
  - Nephrosis
  - Nephritis
- **Post-renal causes:**
  - Cystitis
  - Urethritis
  - Pyetitis

Bence Jones Proteins

- Bence Jones proteins are found in urine in cases of myeloma.

- Their occurrence is extremely uncommon in urine.
- Bence Jones proteins can be detected by the heat coagulation method.

Glucose (Sugars)

- Small amounts of reducing sugars are excreted in urine, which are not detected by the test.
- In abnormal conditions, reducing sugars are excreted in urine.

Methods for Sugar Detection

- Benedict's qualitative test
- Fehling test
- Tests for specific reducing sugars
- Fermentation test: glucose and fructose
- Mucic acid test for lactose and galactose
- Seliwanoff's test for fructose

Ketone Bodies

Acetone, acetoacetic acid, and β-hydroxybutyric acid are collectively known as ketone bodies.

Causes of Ketonuria

- Diabetes mellitus
- Starvation

Method for Ketone Bodies Detection

- Rothera's test

Bile Pigments

- Bile pigments are bilirubin and biliverdin.
- Normal urine does not contain bile pigments.

Methods for Bile Pigments Detection

- Fouchet's test
- Gmelin's test
- Rosenbach test

Bile Salts

- Bile salts are sodium and potassium salts of glycocholates and taurocholates.
- Bile salts are present in urine in obstructive jaundice.

Method for Bile Salts Detection

- Hay's test

Urobilinogen

Urobilinogen is normally present in urine in very small amounts. The excretion of urobilinogen is increased in hemolytic jaundice and infective hepatitis.

Method for Urobilinogen Detection

- Ehrlich's test

Blood

Blood appears in urine in hematuria and hemoglobinuria. Hematuria consists of hemoglobin pigments and unruptured corpuscles.

Method for Blood Detection

- Benzidine test

Chemical Examination of Urine for Normal Constituents

Chloride

- Cl is the main anion in urine.
- It is excreted as sodium chloride.
- The average diet excretes 10–12 g of chloride per day.

Test for Chloride

Principle: When acidified urine reacts with silver nitrate, silver chloride forms a white precipitate.

$$XCl + \text{excess } AgNO_3 \longrightarrow AgCl + AgNO_3$$

$$AgNO_3 + NH_4CNS \longrightarrow \underset{\text{white ppt}}{AgCNS} + NH_4NO_3$$

Test	**Observation**	**Inference**
Take urine (3ml) in a test tube ↓ Add conc. nitric acid (0.5 ml) ↓ 3% silver nitrate (1 ml) ↓ Mix well	A white precipitate is formed.	The presence of chloride in urine sample

Chloride ions decrease in urine because of

- Fasting
- Vomiting and diarrhea
- Diabetes insipidus
- Crushing's syndrome
- Excessive sweating

Chloride ions increase in urine because of

- High protein diet
- Acute hyperthyroidism
- Cystinuria
- Reduction in renal dysfunction

Sulfates

- Sulfates are derived from the metabolism of sulfur-containing amino acids such as cysteine, cystine, and methionine.
- Sulfates are of three forms:
  1. Organic sulfates—ethereal sulfates (5%)
  2. Inorganic sulfates of Na and K (80–85%)
  3. Neutral sulfates (5%)

Test for Inorganic Sulfates

Principle:

$$SO_4^{2-} + BaCl_2 \xrightarrow{HCl} BaSO_4 + KCl$$

Test	Observation	Inference
Take urine (3ml) in a test tube ↓ Add conc. HCl acid (1 ml) ↓ 10% $BaCl_2$ (Dropwise) ↓ Mix well	A white precipitate is formed. 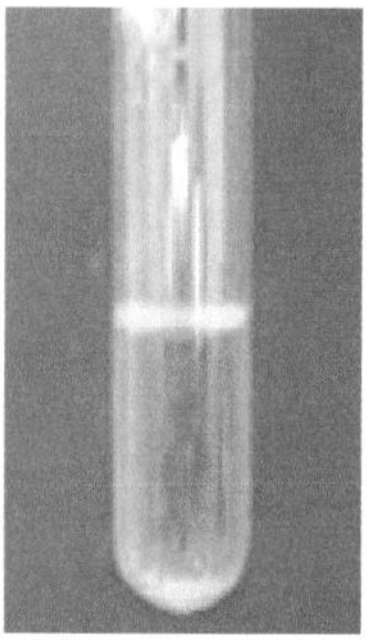	Presence of inorganic sulfates

Inorganic Phosphorus

- Abundant mineral
- The phosphorus of blood is four types:
  1. Inorganic phosphorus present as H_2PO_4 and HPO_4^{2-}
  2. Organic/ester phosphorus
  3. Lipid phosphorus
  4. Residual phosphorus
- Inorganic phosphorus is measured in serum and urine
- Buffer
- 80% present in bones and teeth
- Hydroxyapatite in bone structural support of the body
- Nucleic acid, nucleotides like ATP, GTP, ADP, etc.
- Normal values of inorganic phosphorus in serum:
  Adult = 2.5 to 4.5 mg/dl
  Children = 3.0 to 5.5 mg/dl
  Infant = 4.5 to 8 mg/dl
- The normal value of inorganic phosphorus in urine is 0.5 to 1.5 g/24 h.

Test for Phosphorus: Fike and Subbarow

Principle:

Perchloric acid + Ammonium Molybdate ⟶ Ammonium chloride + Molybdic acid

Molybdic acid + Inorganic Phosphorus ⟶ Phosphomolybdic acid

1,2,4 Amino Naphthol Sulphonic Acid + Phosphomolybdic acid ⟶ Blue Colored Compound

Test	Observation	Inference
Take Protein Free Filtrate (2.5 ml) in a test tube ↓ Add 20% Trichloroacetic acid (2.5 ml) ↓ Ammonium Molybdate (1 ml) ↓ Amino Naphthol Sulfonic Acid (0.2 ml) ↓ Mix well ↓ Keep at RT (20 min.) ↓ Λ_{620}	A blue colored compound is formed. 	Indicates the presence of inorganic phosphates

Hyperphosphatasemia:

- Hypervitaminosis D
- Hypoparathyroidism
- Chronic nephritis

Hypophosphatasemia:

- Hyperparathyroidism
- Rickets/osteomalacia

Calculation:

$$\text{In Urine} = \frac{\text{OD of Test}}{\text{OD of Standard}} \times \frac{\text{Amount of Standard } (\mu g)}{\text{Volume of Sample}} \times \frac{1000}{1000}$$

$$\text{In Serum} = \frac{\text{OD of Test}}{\text{OD of Standard}} \times \frac{\text{Amount of Standard } (\mu g)}{\text{Volume of Sample}} \times \frac{100}{1000}$$

Results = ...mg/L (Urine)

...mg/dl (Serum)

Calcium

- Ca^{2+} is the most abundant mineral in human bodies.
- An average adult body contains 25,000 mmol.
- The total Ca^{2+} present in extracellular fluid is 22.5 mmol (plasma: 9 mmol).
- Most of it exists as complex inorganic hydrated salts (hydroxyapatite, $Ca_{10}(PO_4)_6(OH)_2$).
- Calcium in plasma:
  - 50% in the ionized form (active form)
  - 45% as binding protein (albumin)
  - 5% as anions (citrate, phosphate, sulfate), complexed with a salt
- Normal value of Ca^{2+} in serum = 9–11 mg/dl
- Normal value of Ca^{2+} in urine = 100–300 mg/24 h

Test for Calcium

Principle:

Ca^{2+} + O-Cresolphtalein-OH → Colored Complex

Test	Observation	Inference
Take Urine (3 ml) in a test tube ↓ Add O-Cresolphtalein-OH (2 ml) ↓ Mix well ↓ Λ_{575}	Colored complex is formed.	Indicates the presence of calcium.

Ca^{2+} excretion increases in

- Hyperparathyroidism
- Hyperthyroidism
- Hypervitaminosis D

Calculation:

$$\text{In Urine} = \frac{\text{OD of Test}}{\text{OD of Standard}} \times \frac{\text{Amount of Standard } (\mu g)}{\text{Volume of Sample}} \times \frac{1000}{1000}$$

$$\text{In Serum} = \frac{\text{OD of Test}}{\text{OD of Standard}} \times \frac{\text{Amount of Standard } (\mu g)}{\text{Volume of Sample}} \times \frac{100}{1000}$$

Results = ...mg/L (Urine)

...mg/dl (Serum)

Ammonia

- Ammonia is discharged as an ammonia salt.
- The daily excretion of ammonia is 0.5–0.8 g/day.
- It is made by the kidneys from glutamine and other amino acids.

Test for Ammonia

Principle: Ammonia contained in urine is released by heat.

Due to the generation of alkaline ammonium gas, the color changes from red litmus to blue.

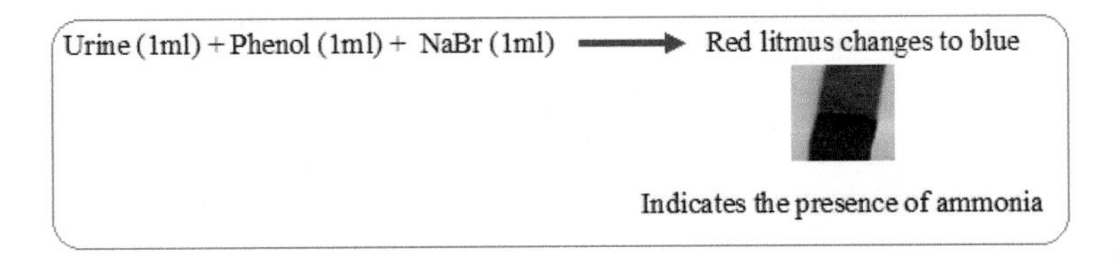

Clinical Significance

Increased Level:

- Diabetic ketoacidosis
- Urinary tract infection

Decreased Level:

- Alkalosis
- Nephritis

Urea

Test for Urea: Sodium Hypochlorite Test

Principle:

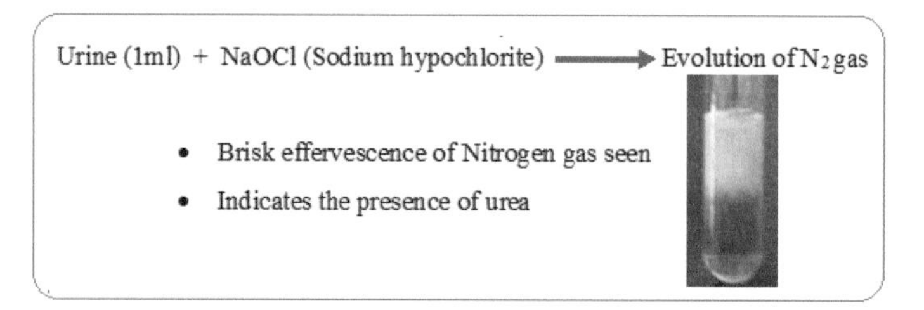

Uric Acid

- Uric acid is the final degradation product of nucleic acids by the catabolism and degradation of human cells that we eat.
- Synthesized in the liver and excreted in the urine as uric acid.
- Normal blood uric acid levels are
  - 3.57 mg/dl for men and
  - 3.6 mg/dl for women.
- Normal urinary excretion of uric acid is 250–750 mg/day.
- Uric acid is insoluble in water, but soluble in alkaline solutions.

Test for Uric Acid (Enzymatic)

Principle:

Uric Acid $+ O_2 + 2H_2O \xrightarrow{\text{Uricase}}$ Allantoin $+ CO_2 + 2H_2O_2$

H_2O_2 + 4AA + ADPS $\xrightarrow{\text{Peroxidase}}$ Quinonemime + $4H_2O$

Pink Color λ_{540nm}

4AA- 4-Aminoantipyrine

ADPS- Aniline derivative, (N-Ethyl-N-(3-sulfopropyl)- 3- methoxy aniline, sodium salt, monohydrate ESPAS

Reduction in Phosphotungstic Acid

Principle:

Uric Acid + Phosphotungstic acid $+ O_2 + 2H_2O \rightarrow$ Allantoin $+ CO_2 + 2H_2O_2$

Allantoin $\xrightarrow[\text{Reduction}]{\text{Alkali}}$ Tungsten Blue Color $\lambda_{700\ nm}$

Test	Observation	Inference
Serum: Protein free filtrate of serum prepared by adding ↓ Serum (0.5 ml) in test tube + H_2SO_4 (3/2 N) + 10% Sodium Tungstate (0.5 ml) ↓ Mix well (Keep for 5 min. at RT) ↓ Centrifuge at 3000 rpm (5 min.) **Urine:** Protein free filtrate of urine prepared by adding ↓ Diluting urine 1:10 with distilled water ↓ Following same process as Serum	Deep blue color is formed.	Presence of uric acid

Procedure:

Reagent in Test Tube	Blank	Standard	T. Serum	T. Urine
Working Standard (uric acid) (μl)	—	50	—	—
Protein free filtrate (serum/urine (μl)	—	—	3 ml	3 ml
Distilled water	3 ml	2.5 ml	—	—
Na_2CO_3	1 ml	1 ml	1 ml	1 ml
Phospotungstic acid (ml)	1 ml	1 ml	1 ml	1 ml

Mix well and store for 20 min.
Read absorbance at 700 nm at room temperature (RT)

Calculation:

$$\text{In Urine} = \frac{\text{OD of Test}}{\text{OD of Standard}} \times \frac{\text{Amount of Standard } (\mu g)}{\text{Volume of Sample}} \times \text{Dilution Factor} \times \frac{1000}{1000}$$

$$\text{In Serum} = \frac{\text{OD of Test}}{\text{OD of Standard}} \times \frac{\text{Amount of Standard } (\mu g)}{\text{Volume of Sample}} \times \frac{100}{1000}$$

Results = ...mg/L (Urine)

...mg/dl (Serum)

Point to be remembered: 3 ml of Protein free filtrate is equivalent to 0.3 ml Serum/Urine sample.

Clinical Significance

Hyperuricemia (>7 mg/dl in males and >6 mg /dl in females):

- Gout
- Purine-rich diet
- Lactic acidosis
- Drugs and poisons
- Toxemia of pregnancy
- Fructose intolerance
- Glycogen storage disease type 1

Hypouricemia (<2 mg/dl):

- Liver disease
- Defective tubular reabsorption (Fanconi syndrome)
- Chemotherapy with azathioprine or 6-mercaptopurine
- Overtreatment with allopurinol
- Enzyme deficiency, e.g., xanthine oxidase

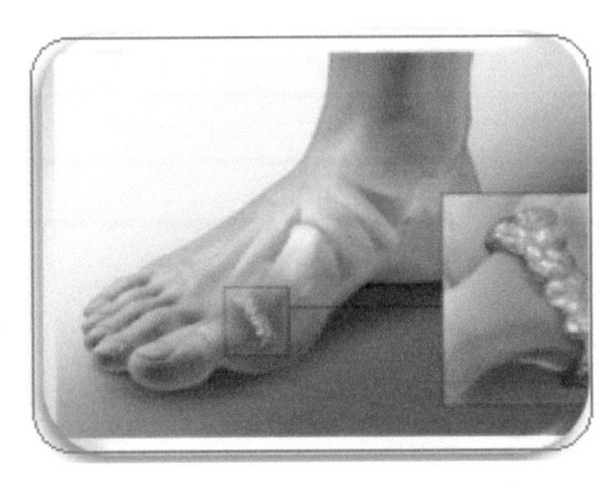

Gout: This is one of the most common clinical symptoms of hyperuricemia. Precipitation of monosodium uric acid from supersaturated fluids causes the clinical signs and symptoms of uric acid deposits to be observed. It is not a single disorder, but a group of disorders that can appear as follows:

- Increased uric acid serum concentration.
- Relapsed seizures of a characteristic type of acute arthritis in which crystals of monosodium urate monohydrate are detected in synovial leukocytes.
- Aggregate deposits of monosodium urate monohydrate crystals, primarily in and around the joints of the extremities: cartilage, synovial bursa, and subcutaneous tissue (tophus).

Creatinine Test (Jaffe's Test)

Principle: Creatinine reacts with picric acid in an alkaline medium to form red–orange creatinine picrate.

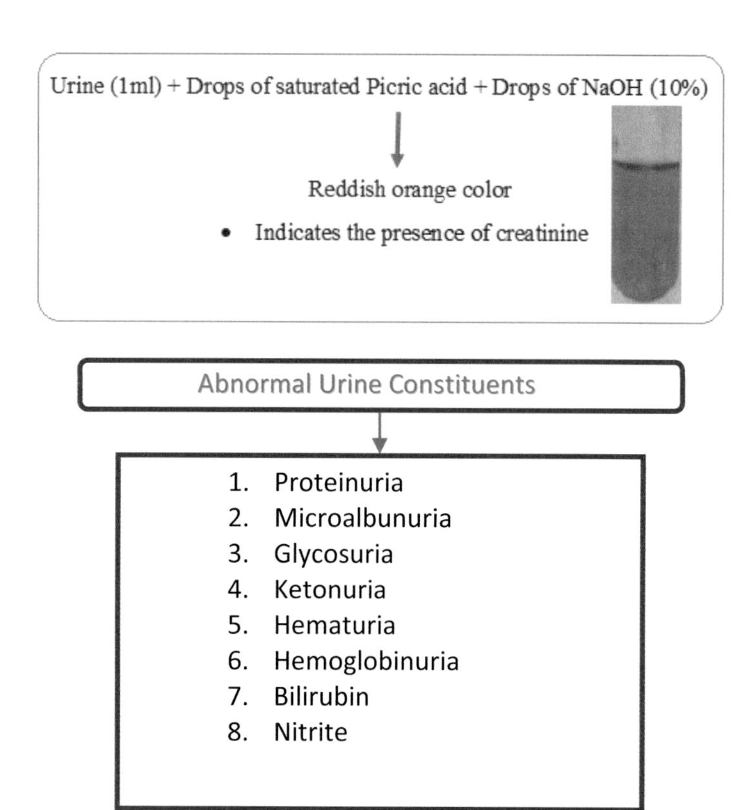

Proteins

Normally, proteins are present in urine only in traces (<150 mg/day), not detectable by routine methods. These consist of albumin, some plasma proteins, and glycoproteins.

- An increased quantity of proteins in urine (i.e., **proteinuria**) can be caused by increased glomerular permeability, reduced tubular reabsorption, increased secretion of proteins from the renal tract, and increased concentration of low-molecular-weight proteins in plasma.
- Sometimes a peculiar protein known as Bence Jones protein (light chain immunoglobulins) is excreted in cases of multiple myeloma.
- These proteins precipitate between 40 and 60°C and redissolve on further heating to 100°C. They reappear again on cooling.

Tests for Proteins

1. Heat Coagulation Test

- Fill two-thirds of a test tube with urine solution, and add 1–2 drops of the chlorophenol red indicator.
- The color of the solution turns pink. Now, heat its upper part; the lower part serves as control.
- Turbidity appears in the upper part, showing the presence of proteins.
- If the color of the solution becomes red instead of pink on adding the indicator, add a few drops of acetic acid to reach the isoelectric pH of albumin (5.6).
- On the other hand, if the developed color is yellow, add a few drops of sodium carbonate.
- Sometimes turbidity may occur due to phosphates, but it dissolves on adding a few drops of acetic acid. But the turbidity due to proteins does not dissolve.

Test	Observation	Inference
Take the solution to be tested up to 3/4th of the test tube ↓ Hold the tube over a flame in a slanting position ↓ Boil the upper portion of the test tube ↓ Lower half serves as control ↓ Add 1% acetic acid drop by drop	A white coagulum is seen at the upper part of the heated test tube.	Indicates coagulable proteins such as albumin

Note:

- When a detectable amount of protein appears in the urine, it is called proteinuria/albuminuria.
- The presence of a detectable amount of protein is characteristic of kidney disease.

2. Sulfosalicylic Acid

Principle: Since sulfosalicylic acid is an alkaloid reagent, it neutralizes positively charged proteins to form a precipitate.

Test	Observation	Inference
Place 3 ml of urine in a test tube. Place 3 mL of urine in a test tube. Now, add 20% sulphosalicylic acid into it drop by drop and note the changes.	A white precipitate is formed. 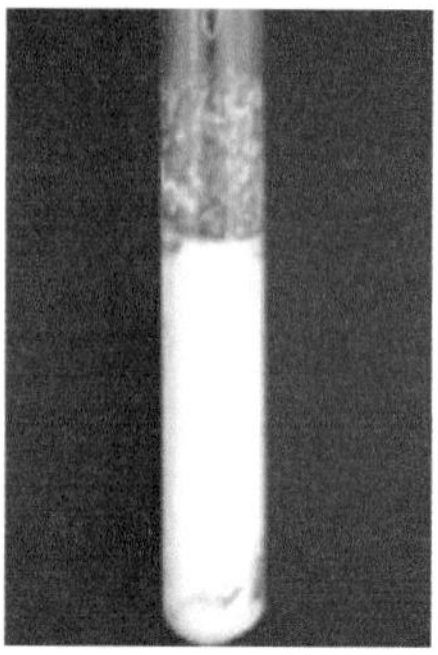	Indicates the presence of proteins

Note:

- This test is used as a routine test for proteins.

3. Heller's Test

Principle: Nitric acid causes precipitation of proteins.

Test	Observation	Inference
Take 3 ml of nitric acid in a test tube. Add 3 ml of urine protein along the sides of test tube.	A white ring is formed at the junction of the two liquids. 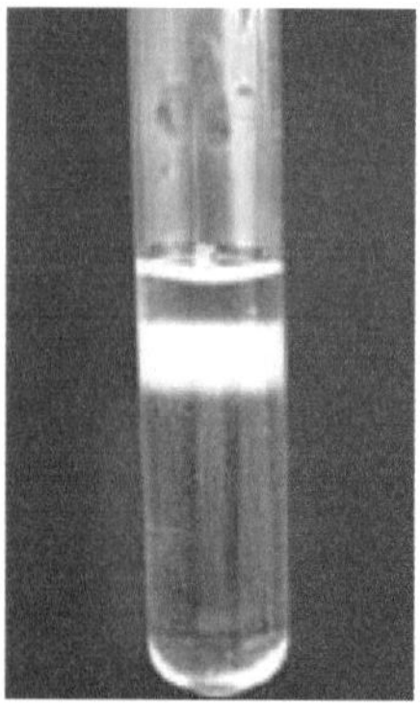	Indicates the presence of proteins

Note:

- This is a highly sensitive test and can be taken as a confirmatory test for protein.
- If urine has a high concentration of urea, urea nitrate may be formed, and it gives a false positive test for proteins.

Carbohydrates

- In normal urine, the amount of reducing sugar is 1–1.5 g/24 h and that of glucose is 500 mg/24 h, which are not detected by routine tests.
- The normal renal threshold for glucose is 180 mg/dl.
- When this level is crossed, glucose starts appearing in urine. This condition is known as glycosuria.
- The presence of detectable amounts of sugar in urine is called glycosuria.
- When the renal threshold for glucose decreases, glucose is present in urine even in the presence of normal blood sugar levels. The condition is termed renal glycosuria. This can be observed during pregnancy.

Common Causes of Glycosuria

- Diabetes mellitus
- Hyperpitutrism
- Hyperthyroidism
- Cushing syndrome
- Pheochromocytoma

Tests for Carbohydrates

1. **Benedict's Test**

   Grading of Benedict's test:

Test	Observation	Inference
Take Benedict's reagent (5 ml) in a test tube ↓ Add given solution (8 drops) ↓ Mix and boil (2 min.) over a small flame ↓ Allow to cool spontaneously.	A brick-red precipitate is formed. 	The given solution is formed reducing sugar.

<table>
<tr>
<td></td>
<td>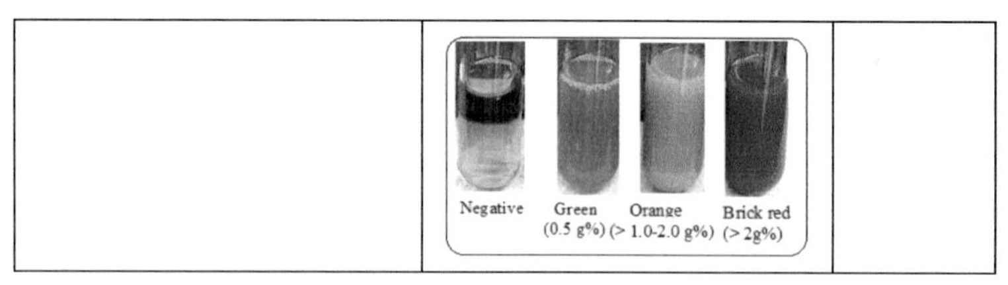
</td>
<td></td>
</tr>
</table>

Note:

- Glucose (glucose): diabetes, renal diabetes
- Fructose (fructose urinary disorder): Fructose metabolic disorder, essential fructosuria, hereditary fructose intolerance
- Galactosemia (galactosemia): Galactosemia
- Lactose (lactose intolerance): Lactose pregnant, lactating women
- Pentose (pentose): Disorders of the uronic acid pathway (essential pentose)

Ketone Bodies

Normally, ketone bodies are not present in urine or present only in undetectable amounts (0.5–3.0 mg in blood). The ketone bodies present are acetoacetate (20%), β-hydroxybutyrate (78%), and acetone (20%). They are produced in liver but utilized extrahepatically to obtain energy. When their concentration increases in blood, the condition is known as ketonemia, and when they are excreted in urine in detectable amounts, the condition is termed ketonuria.

Causes

- Diabetic ketoacidosis
- Starvation
- Severs vomiting
- Glycogen storage disease
- Toxemia of pregnancy
- High fat diet

Test for Ketone Bodies

Rothera's Test
Principle:

$$\text{Acetoacetate} + \text{Sodium nitroprusside} \xrightarrow{\text{Alkaline pH}} \text{Purple}$$

Test	Observation	Inference
Take 5 ml of urine ↓ Saturate it with solid ammonium sulphate ↓ Add 2-3 drops sodium nitroprusside solution ↓ Mix well and add liquid ammonia along the sides of the test tube ↓ A purple ring at the junction of the liquids	A permanganate colored ring is formed.	Indicates the presence of ketone bodies

Note:

- Ketone bodies include acetone, acetoacetic acid, and β-hydroxybutyric acid.
- Rothera's test is very sensitive. It can detect even a small amount of acetone or acetoacetic acid.
- β-Hydroxybutyric acid does not respond to Rothera's or Gerhardt's tests because it does not have a ketone group. Oxidation converts it to acetoacetic acid and then to acetone with positive results.
- The excretion of ketone bodies in urine is called ketonuria. This happens in ketosis.
- Total ketone bodies detected in normal urine are approximately 20 mg/day.

Bile Pigments

Bilirubin is the main bile pigment in the human body. It is formed by the breakdown of heme in reticuloendothelial cells and is released into circulation. It is unconjugated bilirubin and, being water insoluble, requires a carrier for transport, i.e., albumin. On hepatic uptake, it gets conjugated with glucuronic acid and is excreted in bile. Conjugated bilirubin regurgitates into systemic circulation in hepatic and post-hepatic jaundice due to intrahepatic and extrahepatic obstructions, respectively. Being water soluble, it gets excreted in urine. Urine containing bilirubin has a typical beer brown color when voided and produces yellow foam on vigorous shaking. The sample to be analyzed should be fresh; otherwise, bilirubin gets converted to biliverdin.

Test for Bile Pigments

Fouchet's Test

$BaCl_2$ reacts with sulfur radicals in urine to form barium sulfate. If bilirubin is present in urine, it adheres to the precipitate. TCA helps to extract the adsorbed pigments, which are then detected by the oxidation of bilirubin to biliverdin by ferric chloride.

Procedure:

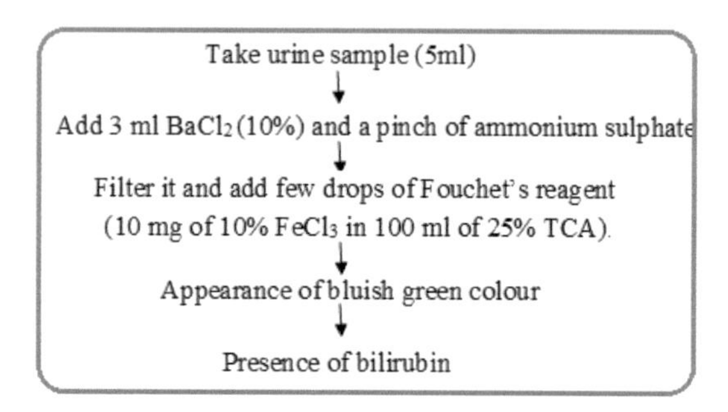

Urobilinogen

Urobilinogen is a colorless compound derived by the bacterial oxidation of bilirubin in intestines, which is reabsorbed by the enterohepatic circulation and excreted in urine. It decreases in the following conditions:

- Obstructive biliary tract disease
- Antibiotic therapy that kills intestinal bacteria

It increases in the following condition:

- Hemolytic anemia

Test for Urobilinogen

Ehrlich's Test

Take 3 ml of urine, and add 3 ml of Ehrlich's reagent (p-dimethylaminobenzaldehyde). Wait for 10 min and add saturated sodium acetate solution dropwise. The appearance of pink color indicates the presence of urobilinogen.

Bile Salts

Bile salts are sodium and potassium salts of glycocholic and taurocholic acids. They are formed in the liver from cholesterol, help in reducing surface tension, and aid in the digestion and absorption of fat. They are excreted in bile but may regurgitate in the systemic circulation in hepatic and post-hepatic jaundice. Being water soluble, they get excreted in urine.

Test for Bile Salts

Hay's Test

Principle: Hay's test is based on the fact that bile salts reduce the surface tension of urine and lower sulfur. Bile salts reduce surface tension, so powdered sulfur sinks in urine samples. Take two test tubes. Fill one of them up to the brim with the urine sample. Fill the other with distilled water to act as the control. Sprinkle sulfur powder in both the tubes. If the sulfur particles sink, bile salts are present in urine. Urine preserved with thymol gives a false positive test.

Test	Observation	Inference
Put 2 ml of a urine sample in a test tube. Sprinkle a little fine sulfur powder over the surface of urine. Observe without mixing.	The sulfur powder sinks to the bottom.	Indicates the presence of bile salts.

Note:

- Bile acids are sodium and potassium salts of glycocholic acid and taurocholic acid.
- Normally, bile salts and bile pigments that do not enter the general circulatory system are not included in normal urine.
- However, when the intrahepatic or post-liver bile flow is obstructed, regurgitation generally occurs and bile acids appear in urine.

- Bile salts are present in urine along with the bile pigments of obstructive jaundice.
- It is not a specific test for bile acids, but it is usually done to check for bile acids.
- Alcohol and salicylate give false positive tests.

Blood

The presence of red blood cells in urine is called hematuria. Blood is usually not present in urine. However, it can be found under the following conditions:

- Malignancy of the kidney, urinary tract, and bladder
- Renal calculi
- Urinary tract infection
- Malignant hypertension
- Sickle cell anemia
- Trauma to the urinary tract

The presence of free hemoglobin in urine is called hemoglobinuria. This free hemoglobin binds with the haptoglobin, so decreased haptoglobin levels can be seen in case of intravascular hemolysis. The presence of myoglobin in urine is called myoglobinuria. It is observed in crush and muscular illness.

Test for Blood

Benzidine Test

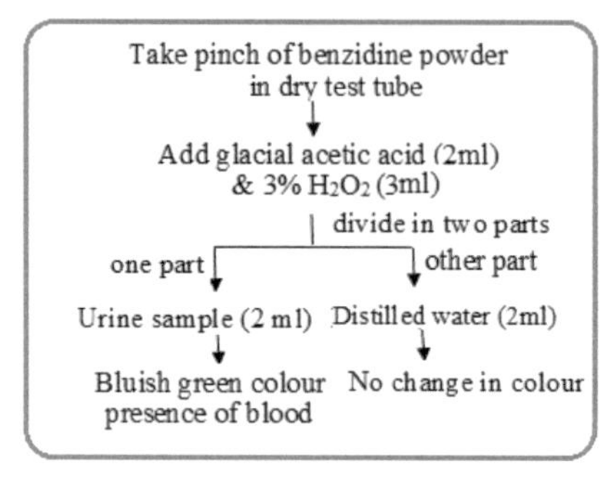

Practical Biochemistry Spots

Spot 1

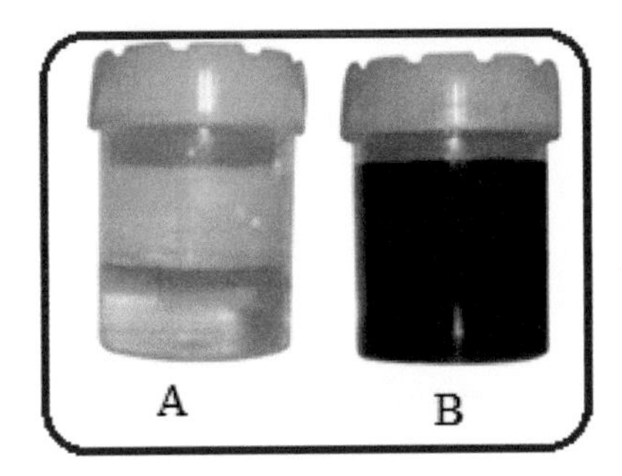

1. Identify the diseases from the colors of the urine samples.
2. Which enzyme is deficient in this state?
3. Name the intermediate of the metabolic pathway accumulated in this disorder.
4. What are the manifestations of this condition?

Spot 2

1. What is the normal volume and appearance of urine?
2. What are the inorganic constituents of (physiological) urine?
3. What are the abnormal constituents of (pathological) urine?
4. What is the cause of decreased urine volume?

Spot 3

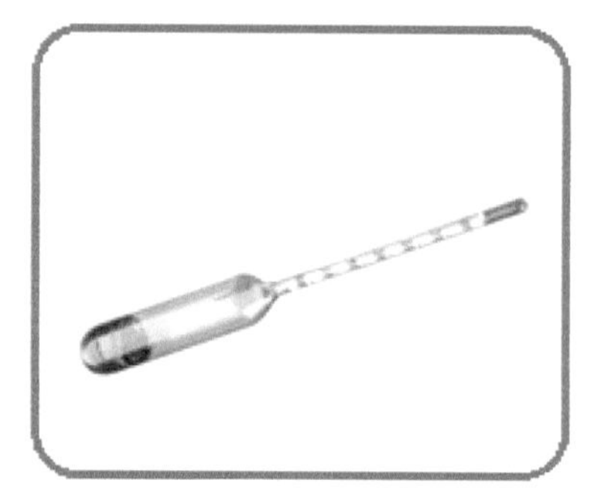

1. Identify the instrument.
2. Mention its use.
3. What is the normal specific gravity of urine?
4. What are the abnormal constituents of urine?

Spot 4

1. Identify the test depicted.
2. What is it positive for?
3. Write the principle of the test.
4. Name a clinical condition in which this test would be positive.

Questions

1. What is the normal specific gravity of urine?
2. What are the common causes of increase in specific gravity?
3. What are the common causes of anuria and oliguria?

Chapter 5

Carbohydrates

Carbohydrates are aldehyde or ketone derivatives of polyhydroxy aldehydes. Glucose is the main form of carbohydrate absorbed from the gut. The main function of carbohydrates is to provide energy. Humans can synthesize glucose from non-carbohydrate sources such as lactates, gluconeogenic amino acids, propionic acid, glycerol, and pyruvate by a process known as gluconeogenesis.

Types of Carbohydrates

Monosaccharides: Monosaccharides cannot be hydrolyzed further. Based on the number of carbon atoms, they can be further divided into troses, tetroses, pentoses, hexoses, and heptoses. Glucose, fructose, and galactose are biologically important monosaccharides in human beings.

Disaccharides: They consist of two monosaccharide units. For example,

- Sucrose = Glucose + Fructose
- Lactose = Glucose + Galactose
- Maltose = Glucose + Glucose

Clinical Biochemistry: A Laboratory Guide
Rooma Devi, Aman Chauhan, Simmi Kharb, and Chandra Shekhar Pundir

ISBN 978-981-4968-75-1 (Hardcover), 978-1-003-45566-0 (eBook)
www.jennystanford.com

Polysaccharides: They are made up of more than 10 monosaccharide subunits. They are of two types: homopolysaccharides and heteropolysaccharides.

- **Homopolysaccharides** are made up of one type of monomeric subunits; for example, cellulose is a polymer of glucose.
- **Heteropolysaccharides** are made up of different types of monomeric subunits; for example, glycosaminoglycans.

Oligosaccharides: They are made up of less than 10 monosaccharide units.

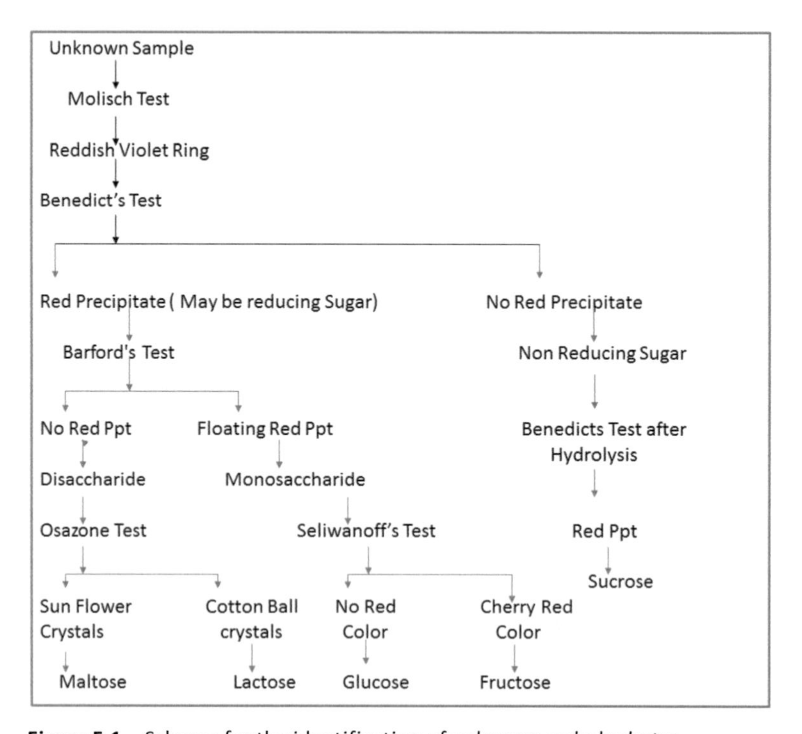

Figure 5.1 Scheme for the identification of unknown carbohydrates.

Qualitative Estimation of Carbohydrates

The various tests for carbohydrates are as follows:

- Molisch's test for carbohydrates

- Iodine test for polysaccharides
- Benedict's test for reducing sugars
- Barfoed's test for monosaccharides
- Seliwanoff's test for keto hexose
- Osazone test for distinguishing reducing sugars based on crystal formation

Physical Properties

- **Color:** Colorless except starch, which is pale white
- **Clarity:** Clear except starch, which is cloudy
- **Odor:** Odorless
- **Reaction to litmus:** Neutral

Chemical Tests

Molisch's Test

Principle: Carbohydrates, when processed with concentrated H_2SO_4, dehydrate to furfural. The hydroxymethylfurfural derivative (chromogens) forms a coloring product by condensation with alpha-naphthol.

$C_5H_{10}O_5$ (Pentose sugar) + H_2SO_4 → Furfural compound + H_2O + H_2SO_4

Furfural compound —(α-naphthol solution + H_2SO_4)→ Purple-Violet colored compound

Test	Observation	Inference
Take 2 ml solution in test tube ↓ Add Molisch reagent (1-2 drops) ↓ Mix. Incline the test tube ↓ overlay conc. H_2SO_4 (2 ml) ↓ along the sides of test tube ↓ Two layers form	At the junction of the two liquids, a violet/purple ring is formed.	• In the given solution, carbohydrates are present. • It is a general test for finding the presence of carbohydrates.

Molisch's test reagent: It is a 5% solution of alpha-naphthol in ethyl alcohol.

Note:

- It is a general test for all carbohydrates.
- This test is given by sugars with at least five carbons because it forms compound derivatives of furfural.
- Impurities in the reagent give a green ring (negative test).
- A green ring occurs even in the absence of carbohydrates due to an excess of alpha-naphthol.

Iodine Test

Principle:

Starch + I_2 ⟶ Starch I_2 ······ I_3^- + Starch ⟶ ⟶ ⟶ (Multi-step Process) I_2+I^-

Dark blue Complex; Amber

Test	Observation	Inference
Take formed 2 ml Blue Color Solution in Test Tube. ↓ Add 2-3 drops of Iodine solution	• No color • Blue color is formed • Reddish brown color is formed.	• Polysaccharides are absent • The given solution is a polysaccharide (starch). • The given solution is a glycogen.

Note:

- This is a specific test for polysaccharides.
- Excess of iodine should be avoided.
- To prevent hydrolysis, the solution must not be highly heated.

Benedict's Test

It is a semi-quantitative test and is the most widely employed test for the detection of reducing sugar in urine. Reducing sugars have a free aldehyde and ketone group, for example glucose, fructose, galactose, and lactose.

Principle: In a mild alkaline medium, reducing sugars undergo tautomerization to form enediols, which reduce Cu^{2+} (cupric ion) to Cu^{+} (cuprous ions). $Cu(OH)_2$ (cuprous hydroxide) is formed. During the heating process, $Cu(OH)_2$ is converted to Cu_2O (cuprous oxide), which gives different shades of colored precipitates depending on the concentration of the sugar.

$$R\text{-}CHO + Cu(citrate)_2^{2-} \longrightarrow R\text{-}COO^- + Cu_2O(s)$$

An aldose + Benedict reagent (Blue Solution) → Carboxylate anion + Brick-red Precipitate

Test	Observation	Inference
Take Benedict's reagent (5 ml) in a test tube ↓ Add given solution (8 drops) ↓ Mix and boil (2 min.) over a small flame ↓ Allow to cool spontaneously.	• Brick-red precipitate is formed. Negative Green (0.5 g%) Orange (> 1.0-2.0 g%) Brick red (> 2g%)	• The given solution is a reducing sugar.

Note:

- The color of the precipitate depends on the concentration of the sugar present.
- It is often used as a screening test for diabetes.
- Positive results appear in the presence of other reducing substances such as ascorbic acid, glutathione, salicylic acid, uric acid, glucuronide, homogentisic acid.
- Automatic reductions can occur, leading to false positives.

Benedict's test for sucrose: Since sucrose is a non-reducing sugar, Benedict's test will never be positive. Therefore, you need to follow the steps given below.

Acid hydrolysis

Principle: Heating in an acidic environment leads to the hydrolysis of the glycosidic bonds present in disaccharides or polysaccharides.

Benedict's Test After Acid Hydrolysis

Test	Observation	Inference
Take Benedict's reagent (5 ml) in a test tube ↓ Add reducing neutralized hydrolysate Monosaccharides solution (8 drops) ↓ Boil 2 min. for Glucose + Fructose	Brick-red precipitate is formed.	On acid hydrolysis, sucrose is converted to reducing sugar. Monosaccharides (glucose + fructose)

Note:

- It is a specific test for reducing sugars (monosaccharides).
- Time for heating is one of the important factors in this reaction.
- If the boiling period exceeds 30 s, disaccharides will be hydrolyzed to monosaccharides and a red-colored precipitate of cuprous oxide will be formed.
- This test differentiates reducing monosaccharides from reducing disaccharides.

Barfoed's Test

This test is for strong reducing monosaccharides.

Principle:

$$(CH_3COO)_2Cu + 2H_2O \longrightarrow 2CH_3COOH + Cu(OH)_2$$

$$Cu(OH)_2 \xrightarrow{\Delta} CuO + H_2O$$

$$\text{D-Glucose} + 2CuO \longrightarrow \text{D-Gluconic acid} + \underset{\text{Cuprous oxide (Red ppt.)}}{Cu_2O \downarrow}$$

Test	Observation	Inference
Take 2 ml of Barfoed's reagent in a test tube ↓ Add 1 ml of given solution ↓ Mix and boil (30 sec.) ↓ Allow it to cool at room temperature	• A floating red precipitate is formed.	• The given solution is a monosaccharide.

Barfoed's reagent: Copper acetate in glacial acetic acid.

Note:

- This test is specific for reducing monosaccharides and for differentiating between monosaccharides and disaccharides.
- Disaccharides give a negative test.
- The time for heating is one of the important factors in this reaction.
- The boiling period should not exceed 2 min. Otherwise, disaccharides will hydrolyze and give a positive test.

Seliwanoff's Test

This test is for ketohexose, i.e., fructose.

Principle:

D-Fructose —(Conc. HCl, $-3H_2O$)→ Hydroxymethyl Furfural + Resorcinol —(Δ)→ Condensation Product (Red)

Test	Observation	Inference
Take Seliwanoff's reagent (3 ml) in a test tube ↓ Add given solution (1 ml) ↓ Boil (30 sec) ↓ Allow to cool at room temperature	• Cherry red color is formed.	• Solution is a ketose sugar (fructose).

Seliwanoff's reagent: Resorcinol (50 mg) in concentrated HCl (3 ml) and diluted with DW (100 ml).

Note:

- This test is specific for ketohexoses only.
- It is useful in differentiating aldohexoses and ketohexoses.
- This test is very sensitive even for 0.1% fructose. In the presence of glucose along with fructose, sensitivity decreases.

Seliwanoff's test for sucrose: In the case of sucrose, the following procedure has to be followed:

Test	Observation	Inference
Take Seliwanoff's reagent (3 ml) in a test tube ↓ Add given solution (1 ml) ↓ Boil (30 sec) ↓ Allow to cool at room temperature	Cherry red color is formed.	Hydrolyzed sucrose contains a keto sugar.

Osazone Test

Many monosaccharides form a characteristic crystal but with a difference in time of formation, melting point, solubility in boiling water, and nature of crystalline structure. This test is also used to identify reducing disaccharides.

Principle: Phenylhydrazine reacts at pH 5 and temperature 100°C with the carbonyl group of reducing sugars to form a soluble phenyl hydrazone, which forms insoluble osazones on further reactions.

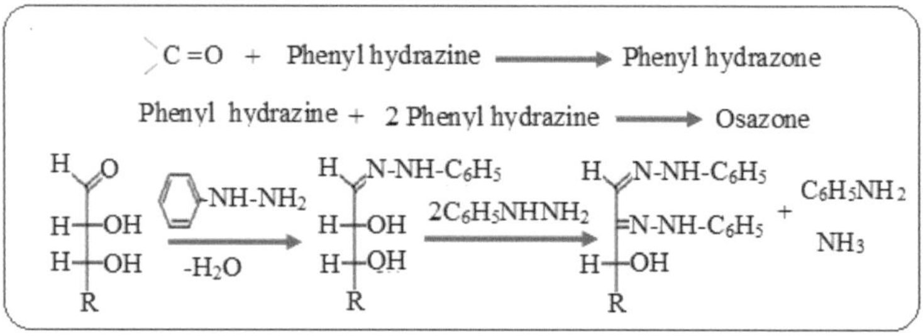

>C=O + Phenyl hydrazine ⟶ Phenyl hydrazone

Phenyl hydrazine + 2 Phenyl hydrazine ⟶ Osazone

H–C=O, H–C–OH, H–C–OH, R + C_6H_5-NH-NH_2 $\xrightarrow{-H_2O}$ H–C=N-NH-C_6H_5, H–C–OH, H–C–OH, R $\xrightarrow{2C_6H_5NHNH_2}$ H–C=N-NH-C_6H_5, C=N-NH-C_6H_5, H–C–OH, R + $C_6H_5NH_2$ + NH_3

Procedure:

Take 5ml solution (glucose, fructose, maltose, lactose) in test tube
↓
Add 0.3 g of Osazone mixture
↓
Add 5 drops of acetic acid
↓
Put all test tubes in water bath

Minimum Time for Formation of Crystals (Osazones)	Appearance of Crystals
Glucosazone (5 min)	• Needle-shaped/broom-stick-shaped appearance 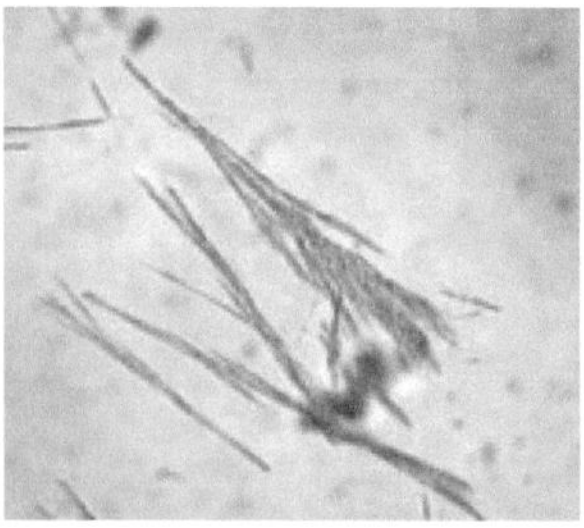

Fructosazone (2 min)	• Needle-shaped/broom-stick-shaped/hay stack appearance 
Lactosazone (30 min)	• Powder-puff-shaped, cotton-ball/badminton-ball-shaped, pincushion with pins/hedgehog-shaped, or flower of touch-me-not plant shaped crystals 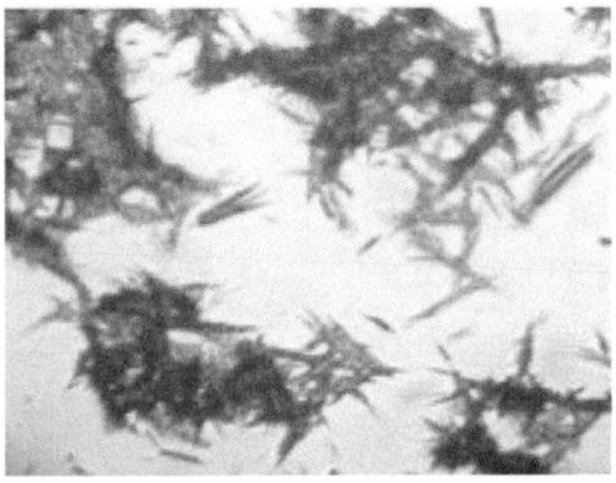
Maltosazone (30–40 min)	• Sunflower-shaped/star-shaped crystals 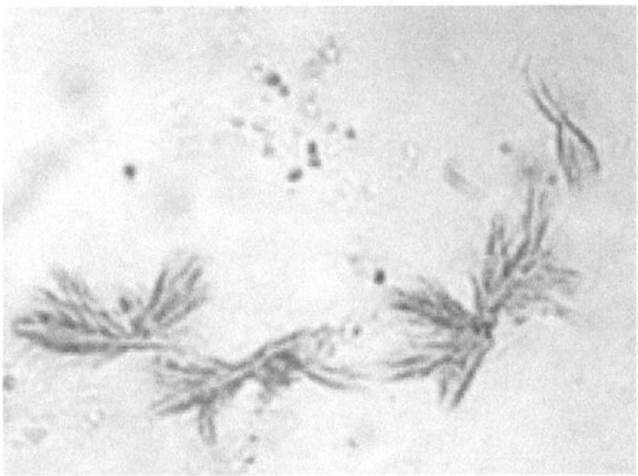

Note:

- There is no free aldehyde/ketone group in sucrose. Hence, sucrose does not produce osazone crystals.
- This test is used to identify the reducing sugar excreted in urine, especially during the period of lactation.
- It is also employed to differentiate glucose and lactose excreted in urine.

Multiple Choice Questions

1. A carbohydrate commonly known as dextrose is
   (a) D-fructose
   (b) D-glucose
   (c) Glycogen
   (d) Dextrin

Ans: b

2. Which of the following is a reducing sugar?
   (a) Sucrose
   (b) Trehalose
   (c) Isomaltose
   (d) Dextrin

Ans: c

3. Which of the following sugars is non-reducing?
   (a) Isomaltose
   (b) Maltose
   (c) Lactose
   (d) Trehalose

Ans: d

4. What is the sugar present in milk called?
   (a) Galactose
   (b) Glucose
   (c) Fructose
   (d) Lactose

Ans: d

5. Benedict's test is not positive for
   (a) Sucrose
   (b) Lactose
   (c) Maltose
   (d) Glucose

Ans: a

6. Seliwanoff's test is positive for
   (a) Fructose
   (b) Lactose

(c) Maltose
(d) Glucose

Ans: a

Practical Biochemistry Spot

Spot 1

1. Name the procedure shown in the picture.
2. What is it used for?
3. What is the principle of this test?
4. List the compound giving this test positive.

Spot 2

1. Name the procedure shown in the picture.
2. What is it used for?
3. What is the principle of this test?
4. List the compound giving this test positive.

Spot 3

1. Name the procedure shown in the picture.
2. What is it used for?
3. What is the principle of this test?
4. What is the interpretation of results based on the color produced?

Questions

1. Write the names of the hydrolytic products of three disaccharides.

Hydrolysis products

a. Maltose (malt sugar) → D-glucose + D-glucose

b. Lactose (milk sugar) → One mole of D-glucose + One mole of D-galactose

c. Sucrose (table sugar) → One molecule of D-glucose + one mole of D-fructose

2. Name the non-reducing disaccharide.

Sucrose is a non-reducing disaccharide. In sucrose, both the aldehyde group of glucose and the ketone group of fructose are bound by an $\alpha 1 \rightarrow 2$ bond. Therefore, free aldehyde or ketone groups are not available.

3. Differentiate between dextrins and dextrans.

Dextrins	Dextrans
Are hydrolytic	Synthetic polymer of D-glucose products of starch
Used in infant feeding	Used as a plasma expander; when given intravenously in hemorrhage (blood loss), it increases the blood volume.

Chapter 6

Blood Glucose

Glucose, the major carbohydrate in blood, is also the major source of energy for the body's cells. Blood sugar includes all hexoses, even though the main hexose present in the blood is glucose. Here, blood sugar means blood glucose. It is an essential nutrient for the brain and red blood cells (RBCs). The most common disorder of glucose metabolism is **diabetes mellitus**, a disease in which the blood glucose is elevated because of the lack of insulin regulation.

The state of having elevated blood glucose is called **hyperglycemia**. Decreased blood glucose is called **hypoglycemia**, which is a less common disorder of glucose metabolism. The glucose test is most often used to aid in diagnosing and managing diabetes and hypoglycemia.

Blood Glucose Regulation

The body has many metabolic pathways that utilize glucose, and several hormones work together to keep blood glucose within a fairly narrow range.

- **Insulin:** Insulin *lowers* blood glucose by increasing the cellular uptake of glucose and increasing the rate of glycolysis. Insulin also increases the rate of conversion of glucose to glycogen.

Clinical Biochemistry: A Laboratory Guide
Rooma Devi, Aman Chauhan, Simmi Kharb, and Chandra Shekhar Pundir

ISBN 978-981-4968-75-1 (Hardcover), 978-1-003-45566-0 (eBook)
www.jennystanford.com

- **Glucagon and other hormones:** Hormones such as growth hormone, epinephrine, somatostatin, cortisol, and **glucagon** act in a variety of ways to *increase* blood glucose concentration. These hormones are sometimes called insulin antagonists because their action is opposite to the action of insulin. Together, insulin and glucagon are the major hormones that maintain blood glucose concentrations in the normal range.

Broadly two methods are available for blood glucose estimation:

1. Reduction (chemical) methods
2. Enzymatic methods

Reduction (Chemical) Methods

A. Oxidation–Reduction

Glucose + Alkaline copper tartrate $\xrightarrow{\text{Reduction}}$ Cuprous oxide
Alkaline copper reduction

Folin-Wu method:
$CuSO_4$ + Phosphomolybdic acid $\xrightarrow{\text{Oxidation}}$ Molybdenum
Blue color end product

Benedict's method: Modification of the Folin-Wu method for qualitative urine glucose
The reagent contains sodium citrate and sodium carbonate with Cu_2SO_4. It gives color according to the concentration of the glucose (green...yellow... brown...red)

Nelson–Somogyi method:
Cu^{++} + Arsenomolybdic acid $\xrightarrow{\text{Oxidation}}$ Arsenomolybdenum
Blue end product

B. Condensation

Ortho-toluidine method: Use aromatic amines and hot acetic acid

- Forms glycosylamine and Schiff base, which is emerald green in color.
- This is the most precise technique, but the reagent used is poisonous.

O-toluidine + Glucose (aldehyde) $\xrightarrow{\text{Heat/Acidity}}$ **Glucosamine (colored)**

Enzymatic Methods

A. Glucose oxidase

$$\text{Glucose} + O_2 \xrightarrow[\text{Oxidation}]{\text{Glucose oxidase}} \text{Cuprous oxide}$$

Saifer–Gerstenfeld method: $H_2O_2 + \text{O-dianisidine} \xrightarrow[\text{oxidation}]{\text{Peroxidase}} H_2O$ Oxidized chromogen

- Inhibited by reducing substances such as BUA, bilirubin, gluthione, and ascorbic acid

Kodak Ektachem:

- A dry chemistry method
- Uses reflectance spectrophotometry to measure the intensity of color through a lower transparent film

Glucometer:

- Home monitoring blood glucose assay approach
- Uses a strip immobilized with glucose oxidase

B. Hexokinase

$$\text{Glucose} + \text{ATP} \xleftrightarrow{\text{Hexokinase} + Mg^{++}} \text{Glucose-6-phosphate} + \text{ADP}$$

$$\text{Glucose-6-phosphate} + NADP^+ \xleftrightarrow{\text{G-6PD}} \text{G-Phosphogluconate} + \text{NADPH} + H^+$$

- NADP H as cofactor
- Reduced (NADPH product) is measured in 340 nm.
- More specific than glucose oxidase method due to glucose-6-phosphate, which inhibits interfering substances except when sample is hemolyzed.

O-Toluidine Method

Sample collection for glucose analysis:

- 2 ml of venous blood is collected aseptically in a vial containing a mixture of sodium fluoride and potassium oxalate in a ratio of 1:1.
- NaF inhibits glycolysis by inhibiting the enzyme enolase, and if sample is analyzed after a few hours, the RBCs cannot use glucose and the values remain same.
- Potassium oxalate acts as an anticoagulant.

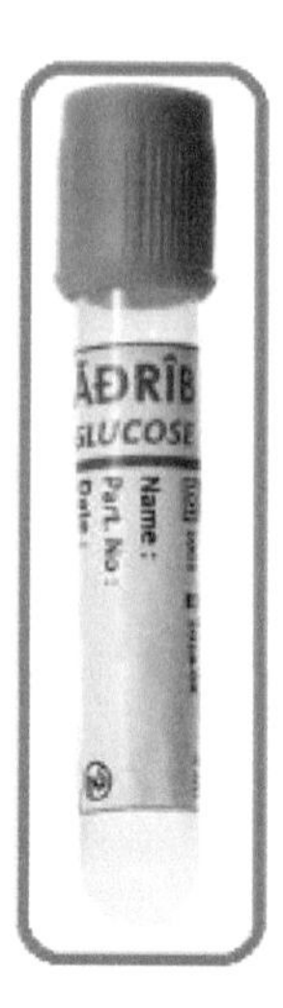

Principle: Ortho-toluidine reacts quantitatively with the aldehyde group of hexoses to form a glycosylamine and Schiff base. The absorbance of this is measured at 630 nm.

$$CH_2OH{-}(CHOH)_4{-}CHO \;(\text{glucose}) + \text{o-toluidine} \rightleftharpoons CH_2OH{-}(CHOH)_4{-}CH{=}N{-}C_6H_4{-}CH_3 \;(\text{blue-green complex})$$

- Galactose can react, and its concentration in blood is negligible.

Preparation of protein free filtrate:

- Protein free filtrate (PFF) is prepared by adding 1.8 ml of 4% TCA and 0.2 ml blood/plasma/serum.
- Mix well and wait for 3–4 min.
- Centrifuge for 10 min.
- Since proteins interfere with the estimation, these are removed first.

Take three test tubes and label them as test, standard, and blank.

	Test	**Standard**	**Blank**
Distilled water	—	—	0.5 ml
Working standard (10 mg%)	—	0.5 ml	—
PFF	0.5 ml	—	—
O-toluidine reagent	2.5 ml	2.5 ml	2.5 ml

- Mix and keep the test tubes in a boiling water bath for 10 min.
- Cool the test tubes under tap water and take OD at 630 nm.

Calculation:
A(Sample)/A(Standard) × 100 (Standard conc.) = mg/dL glucose in the sample

Conversion factor: mg/dL × 0.0555 = mmol/L

$$\frac{\text{OD of Test}}{\text{OD of Stand.}} \times \frac{\text{Conc. of Stand.}}{\text{Vol. of Test}} \times 100 = ...\text{mg\%}$$

Normal range: 60–100 mg%
The O-toluidine reagent is prepared by adding glacial acetic acid and thiourea to ortho-toluidine dye. Thiourea helps to stabilize the color of the reaction product.

Merits:

- It is a simple and rapid method.
- The values obtained provide true glucose levels.

Demerits:

- As ortho-toluidine is a carcinogenic substance, it is dangerous to handle this chemical.

Precautions:

- Ortho-toluidine should be pipetted very carefully.
- An antiglycolytic substance, i.e., NaF, should be added; otherwise, blood glucose levels decrease by 5–7% in 1 h.

Normal range: Fasting blood glucose is 70–110 mg/dl.

Other Methods

Based on reducing properties of glucose

a. ***Folin-Wu method***

Principle: Glucose reacts with copper sulfate in an alkaline medium to form cuprous oxide, which forms a blue complex with phosphotungstic acid, whose intensity is proportional with the amount of glucose in blood.

$$CuSO_4 + \text{Phosphomolybdic acid} \xrightarrow{\text{Oxidation}} \text{Molybdenum (Blue color end product)}$$

Disadvantage:

- The blood glucose level obtained by this method is higher by 20–30 mg% because of the interference of other reducing substances, such as glutathione, vitamin C, and uric acid.
- Lacks specificity.
- The normal range for fasting sample is 90–120 mg%.

b. *Nelson and Somogyi method*

Principle: In this method, heavy metal salts, such as zinc sulfate and barium hydroxide, are used to prepare PFF, which removes non-glucose reducing substances more effectively.
The reagent used is arsenomolybdic acid, and OD is read at 650 nm.
It gives better values than Folin-Wu but is not in common use.
The normal range for fasting sample is 65–110 mg.

c. *Glucose oxidase–peroxidase method*

Principle:

$$\text{Glucose} + O_2 + H_2O \xrightarrow{\text{Glucose Oxidase}} \text{Gluconic Acid} + H_2O_2$$

$$H_2O_2 + \text{Phenol} + 4\ \text{Aminoantipyrin} \xrightarrow{\text{Peroxidase}} \text{Quinonimine} + H_2O$$

Advantages:

- True glucose level is obtained by this method as it is quite specific.
- The color developed is stable.
- Only 10–15 µl of sample is used, which can be taken from the capillaries.

Disadvantages:

- It is expensive.
- Various substances such as uric acid, fluoride, bilirubin, glutathione, and ascorbic acid inhibit the reaction.

d. *Hexokinase method*

Principle: Hexokinase converts glucose to glucose-6-phosphate, which is then oxidized to 6-phosphogluconate by glucose-6-phosphate dehydrogenase in the presence of NADP. The NADPH so formed is measured at 340 nm.

Advantages:

- It can be done manually, by automation, or by colorimetry.
- The values obtained are similar to those obtained by the glucose oxidase method.
- No interference with oxalate, fluoride, or EDTA is observed.

Disadvantages:

- Enzyme preparation needs storage at 5°C in desiccation and gets unstable at 26°C after 5 h.
- The normal range for fasting sample is 60–90 mg%.

Clinical significance:

Causes of hyperglycemia:
Can be divided into

- Endocrinal
- Non-endocrinal

Endocrinal:

- Insulin insufficiency
- Hyperthyroidism
- Hyperpituitarism
- Hyperactivity of adrenal gland
- Excessive glucagon secretion
- Emotional stress

Non-endocrinal:

- Pancreatitis
- Terminal stages of many diseases
- After general anesthesia
- Old age

Causes of hypoglycemia:

- Sugar < 40 mg%
- Insulin overdose
- Pancreatic islet cell tumor and other non-pancreatic tumors
- Adrenocortical insufficiency
- Hypopituitarism
- Primary renal disease leading to renal glycosuria/end-stage renal failure
- Severe exercise
- Alcoholic patients
- Septicemia
- Hepatic failure

Significance of Glucose Estimation

Glucose estimation is of great significance in patients suffering with diabetes mellitus, which can be detected if

- Classical symptoms are present along with random plasma glucose concentration ≥ 200 mg%.
- Fasting plasma glucose ≥ 126 mg%.
- Two-hour postprandial plasma glucose ≥ 200 mg% during OGTT.

A patient is said to have impaired glucose tolerance if

- Fasting plasma glucose < 126 mg%.
- Two-hour OGTT plasma glucose concentration is between 140 and 199 mg%.

Table 6.1 Diabetic profile

Name of Investigation	Method Name	Biological References
B. Sugar (F)	GOD–POD	60–100 mg/dl
B. Sugar (PP)	GOD–POD	70–140 mg/dl
B. Sugar (R)	GOD–POD	70–140 mg/dl
Hb1c	Latex agglutination inhibition	4–6%

Diagnostic Tests for Diabetes and Hypoglycemia

The following tests are used to aid in the diagnosis of diabetes or hypoglycemia:

- Fasting blood glucose
- 2 h postprandial glucose test
- Oral glucose tolerance test (OGTT)
- Hemoglobin A1c test
- Clinical classification of blood sugar

Fasting blood sugar: Blood sample is collected after overnight fasting of 8 h to 12 h.

Postprandial blood sugar: Blood is collected 2 h after normal diet, and blood is collected 2 h after 75 g glucose load.

Random blood sugar: Blood is collected anytime without prior preparation of the patient.

Glucose Tolerance Test

The ability of a person to metabolize a given load of glucose is referred to as glucose tolerance. The glucose tolerance test is a well-standardized test. It is useful in doubtful cases when a patient has symptoms suggestive of diabetes mellitus, but blood sugar value is inconclusive. This test evaluates the degree of tolerance of an individual to glucose load under standard conditions.

Types

1. Oral glucose tolerance test
2. Intravenous glucose tolerance test

Indications

- Patients have symptoms suggestive of diabetes mellitus, but fasting blood sugar is between 100 and 126 mg/dl.
- To rule out benign renal glucosuria.

- During pregnancy, excessive weight gain is noticed, with a history of big baby (more than 4 kg) or miscarriage.

Contraindications

- There is no indication for doing OGTT in a person with confirmed diabetes mellitus.
- Glucose tolerance test has no role in the follow-up of diabetes.
- This test should not be done in acutely ill patients.

Preparation of patients for OGTT

- Overnight fasting should be done (at least 10–14 hours).
- All drugs should be stopped for at least 2 days prior to the test.
- Smoking should be stopped.

Procedure

- A fasting blood glucose sample is drawn in the morning and tested. A corresponding urine sample is also collected and tested.
- The patient is asked to consume a glucose solution (75 g anhydrous glucose in 250–300 ml of water or 1.75 g/kg body weight; maximum 75 g for children).
- Collect five samples of venous blood and urine at half hourly intervals.
- Determine the blood glucose by the specific method, e.g., GOD-POD method.
- Determine urine glucose by a semiquantitative method—Benedict's test.
- Prepare a glucose tolerance curve (blood glucose level – time)

Interpretation

Normal curve:

- The initial fasting glucose is within normal range.
- The maximum peak value is reached within 1 h.

- It returns to the normal fasting level within 2–2.5 h.
- The highest value does not exceed the renal threshold (180 mg%)
- No glucose or ketone bodies are detected in any specimen of urine.

Diabetic curve:

- Fasting blood glucose is raised above 110 mg%.
- The highest value exceeds the renal threshold.
- The highest value is usually reached within 1 h to 1.5 h and crossed the renal threshold.
- The blood glucose level does not return to the fasting level within 2.5 h.
- The urine sample always contains glucose except in some chronic diabetes or nephritis, which might have raised the renal threshold.
- Glycosuria is often seen.

Renal glycosuria curve:

- The curve is normal.
- Due to the lowering of the renal threshold, glucose appears in the urine sample.
- Patients who show no glucosuria when fasting might have glucosuria when blood glucose is raised.
- The renal threshold may be seen in renal disease, pregnancy, and early diabetes.
- It is useful in recognizing borderline cases of diabetes.

Lag curve:

- Fasting blood glucose level is normal.
- Due to rapid absorption, the maximum level is reached in 30 min, which crosses the renal threshold.
- The urine sample shows glucose.
- The return to normal value is rapid and complete.
- The curve is seen in hyperthyroidism, gastroenterectomy, and pregnancy.

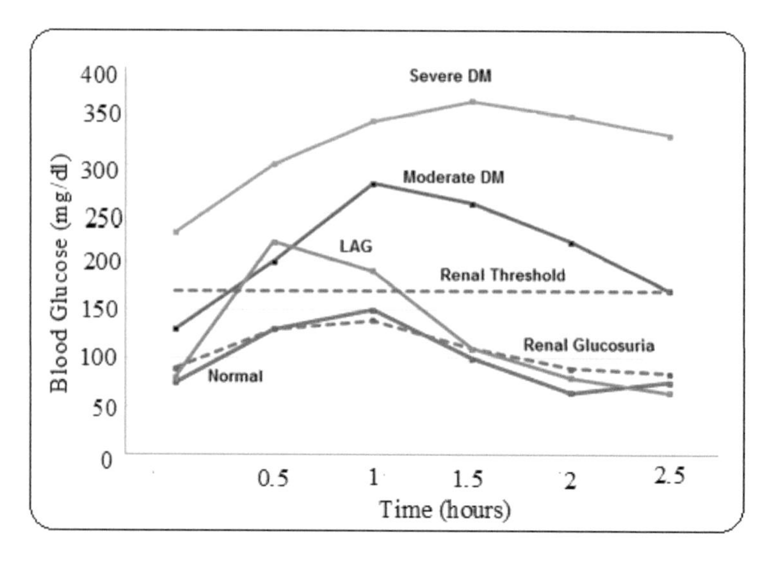

Figure 6.1 Oral glucose tolerance test curve.

Intravenous Glucose Tolerance Test

- This test is performed in patients who cannot tolerate total oral glucose, such as patients of severe nausea, malabsorption syndrome, and vomiting.
- Samples are taken at 10 min intervals for the next hour in these cases.

Glucose Reference Values

The blood glucose reference range is the laboratory's acceptable range of glucose values obtained from statistical analysis of glucose values in the general population.

Normal blood sugar levels:

- Fasting blood sugar level: 70–110 mg/dl
- Postprandial blood sugar level (after 2 h of meal): 70–140 mg/dl

Blood sugar levels in prediabetics (impaired glucose tolerance):

- Fasting blood sugar level: 111–125 mg/dl
- Postprandial blood sugar level: 141–199 mg/dl

Blood sugar levels in diabetics:

- Fasting blood sugar level: 126 mg/dl or more

Diabetes Management

Good diabetes management is important for preventing complications of diabetes caused by microvascular damage.

Hemoglobin A1c

The ability of patients to keep blood glucose levels within acceptable ranges can be assessed by the periodic measurement of hemoglobin A1c (HbA1c). It is recommended that HbA1c be measured in patients with diabetes every 3 or 6 months. HbA1c at or below 7% indicates adequate glucose control.

Recent studies have also shown that HbA1c is the single best test for evaluating the risk of damage to nerves and the small blood vessels of the eyes and kidneys. This microvascular damage leads to the complications of diabetes, such as blindness and kidney failure. Clinical trials have shown that reducing the HbA1c level in people with diabetes and maintaining it below 7% will prevent the development or further progression of complications from diabetes.

An extremely high glucose level can lead to diabetic coma and requires immediate treatment to reduce the glucose level.

Blood glucose exceeding 400 mg/dl is considered dangerously high. Symptoms of extreme hyperglycemia include confusion, lethargy, extreme thirst, weak pulse, dry skin, and nausea.

Normal values:

- Glycohemoglobin A1c: 4.5–5.7%
- Total glycohemoglobin: 5.3–7.5%

Case Study: Diabetes Mellitus

Case Study 1

A 47-year-old businessman was happy that he had lost 8 kg in the last 3 months. He felt he was losing weight as he had started drinking more water than usual though he kept feeling hungry all the time. Getting up at night to empty his bladder was disturbing his sleep and made him feel tired all through the day. His physical examination and lab investigations carried out as part of the yearly health check-up showed the following significant findings:

- BMI: 28
- Fasting plasma sugar: 180 mg/dl
- Urine sugar: absent
- Urine ketones: absent
- Postprandial plasma sugar: 230 mg/dl
- HbA1c: 7.9%

He was asked to follow up with a physician.

1. What is your diagnosis?
2. What other investigations would you like to do?
3. What are the alterations in normal biochemistry that explain the given clinical presentation?
4. What is the significance of raised HbA1c and high BMI in a patient of DM?
5. Why should the blood sample for glucose be collected in a fluoride-EDTA bulb or tube (grey)?

Case Study 2

A 35-year-old software engineer comes to the medicine OPD with the chief complaints of fatigue, weight loss, increase in appetite, thirst, and increased frequency of urination. He gives a history of 2 kg weight loss in 1 month. How do you explain the symptoms and which investigations should be done in this case?

1. What is your diagnosis?
2. What other investigations would you like to do?
3. What is the management of this patient?

Case Study 3

A 46-year-old obese female visited a biochemistry laboratory for routine blood glucose testing. The test reports are as follows:

- Blood glucose (fasting): 142 mg/dl
- Blood glucose (PP): 187 mg/dl
- Urine glucose (fasting): absent
- Urine glucose (PP): present

1. What is your diagnosis?
2. What other investigations would you like to do?
3. What is the management of this patient?

Chapter 7

Proteins

Proteins are defined as sequence-specific polymers of Lα amino acids linked by peptide bonds. They are complex organic substances consisting of carbon, hydrogen, oxygen, and nitrogen. Some proteins also contain sulfur and phosphorus.

Classification

1. **Simple proteins:** Consist of amino acids only. Examples: albumin, globulin
2. **Conjugated proteins:** Conjugated with amino acids and non-protein moieties known as prosthetic groups. Example:
   - Glycoprotein: immunoglobulin
   - Lipoprotein: HDL, LDL, VLDL
   - Metalloproteins: hemoglobin, cytochromes
   - Nucleoproteins: histones
   - Phosphoprotein: milk casein
3. **Derived proteins:** Derived from naturally occurring protein. There are two types:
   a. **Primary proteins:** Formed by agents such as heat, acids, and alkalis that cause only minor changes. Proteins without cleavage by the hydrolysis of peptide bonds and their properties, such as metaproteins.

Clinical Biochemistry: A Laboratory Guide
Rooma Devi, Aman Chauhan, Simmi Kharb, and Chandra Shekhar Pundir

ISBN 978-981-4968-75-1 (Hardcover), 978-1-003-45566-0 (eBook)
www.jennystanford.com

b. **Secondary proteins:** Result from progressive hydrolysis cleavage of the protein's peptide bond, e.g., proteose, peptides, and peptones.

Reactions of Proteins

Reactions of proteins are studied as precipitation reactions and color reactions.

Precipitation Reactions

Proteins are large molecules with variable sizes, shapes, and charges. Protein solubility depends on the distribution of polar hydrophilic groups and nonpolar hydrophobic groups in the molecule. Proteins form colloidal systems in aqueous media. A colloid is a system in which the particles have diameters in the range of 1–200 millimicron. The stability of a protein in a solution will depend mainly on the charge and the degree of hydration (shell of water molecules around the particles). Polar groups of the protein ($-NH_2$, COO–, OH–) tend to attract water molecules around them to form a shell of hydration.

Colloids are two types: suspensoids and emulsoids.

- ***Suspensoids*** are stabilized by the electrical charges over the surface of the molecule.
- ***Emulsoids*** are stabilized by an electric charge over the surface of the molecule and hydration shell around the molecule.

Proteins can be precipitated by:

- Removing their shell of hydration.
- Neutralizing electrical charges.
- Denaturation (disorganization of native protein, loss of biological activity).
- Bringing them to isoelectric pH.

Any factor that neutralizes the charge or removes the shell of hydration will cause precipitation of proteins.

Importance of precipitation of proteins: Precipitation is used to separate proteins from biological fluids such as blood, plasma, and CSF before the estimation of important chemical constituents such as urea, sugar, and creatinine, because proteins interfere in their estimation.

Precipitation by salts

Principle: The addition of neutral salts such as ammonium sulfate leads to the adsorption of the hydration shell along with the neutralization of surface charges leading to protein precipitation. This is known as “salting out.”

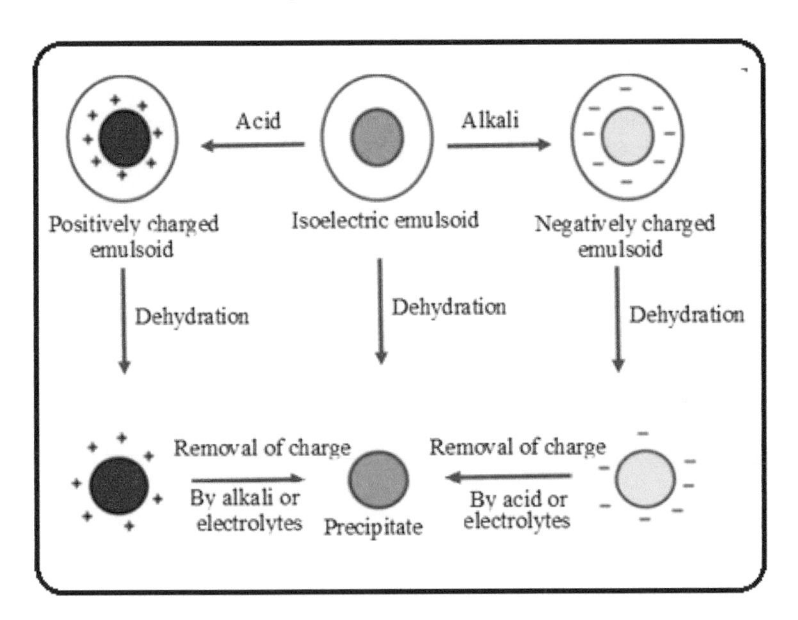

a. Half-saturation with ammonium sulfate

Test	Observation	Inference
3 ml of Protein solution in a test tube ↓ Add equal vol. sat Ammonium Sulfate solution ↓ Shake & Stand for 5 min. ↓ Filter and perform Biuret Test • Biuret test: 40% NaOH (3 ml) & 1% $CuSO_4$ (2-3 drops)	Violet color is formed in the case of albumin, and no violet color is formed in the case of globulin.	The protein contained is albumin.

b. Full saturation with ammonium sulfate

Test	Observation	Inference
3 ml of Protein solution in a test tube ↓ Add equal vol. sat Ammonium Sulfate solution ↓ Shake & Stand for 5 min. ↓ Filter and perform Biuret Test • Biuret test: 40% NaOH (3 ml) & 1% $CuSO_4$ (2-3 drops)	No change in the case of albumin and globulin	Albumin and globulin are not present.

Note:

- A filtrate that does not have any trace of protein gives blue color with Biuret reagent.
- Casein and gelatin are precipitated by half-saturation because they have high molecular weight.

Isoelectric precipitation (casein)

Principle:

$$\underset{NH_3^+}{\text{R-CH-COOH}} \xleftarrow{CH_2\text{-}COOH} \underset{NH_3^+}{\text{R-CH-COO}^-} \underset{KOH}{\overset{CH_2\text{-}COOH}{\rightleftarrows}} \underset{NH_2}{\text{R-CH-COO}^-}$$

PH < I.E.P Soluble — I.E.P. PH=4.5-4.8 ppt. — PH > I.E.P Soluble

The solubility of proteins is minimum at their isoelectric pH as the protein molecules become electrically neutral at this pH.

Test	Observation	Inference
Casein solution (3 ml) in a test tube. ↓ Add 3 drops of bromocresol green indicator ↓ Mix add 1% acetic acid (drop by drop) ↓ Light green color is obtained ↓ Indicates the pH is close to 4.6.	Curdy green precipitate is formed.	Casein is present.

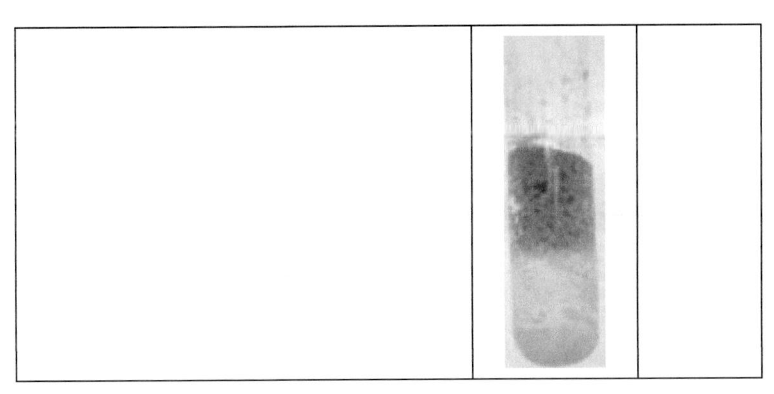

Note:

- At isoelectric pH,
  - The net charge is zero.
  - No mobility in an electric field.
  - Least soluble.
  - The buffer capacity and viscosity are minimal.
  - Maximum precipitation.
- The pH range of bromocresol green is 4–4.6.
- The isoelectric pH of casein is 4.6.
- The isoelectric pH of human albumin is 4.7.
- Proteins have minimum solubility at the isoelectric point.
- The isoelectric point of albumin is 4.7.

  At this pH, the color of the solution is dark green. If the color is not brought to light green (i.e., pH 4.6 of casein), albumin might form a curdy dark green precipitate at pH 4.7. So, while performing the isoelectric precipitation test for casein, bring the pH of the solution to 4.6 (light green) to avoid interference of albumin.

Precipitation with organic solvents

Principle:

- Polar organic solvents such as methanol, ethanol, and acetone reduce the solubility of proteins.
- Organic solvents reduce the hydration of proteins in solution by reducing the number of hydrogen bonds.
- The dielectric constant of the medium also decreases, causing aggregation and precipitation.

Test	Observation	Inference
Protein Solution (1 ml) in a test tube ↓ Add Ethanol (2 ml) ↓ Shake & Stand for 5 min	A white precipitate is formed.	Protein is precipitated by organic solvents.

Note:

- For the precipitation of protein by alcohol, the protein should be in electrolyte form. This may be achieved by dissolving the protein in saline.

Precipitation by alkaloidal reagents

Principle: The negatively charged ions of the alkaloids neutralize the positive charge on the protein causing denaturation, which results in precipitation.

Test	Observation	Inference
Protein Solution (2 ml) in a test tube ↓ Add Picric Acid (drop by drop) ↓ Shake & Stand for 5 min	A thick yellow precipitate is formed.	Protein is precipitated by the alkaloidal reagent.

Protein Solution (2 ml) in a test tube ↓ Add Trichloroacetic Acid (drop by drop) ↓ Shake & Stand for 5 min	A white precipitate is formed. 	Protein is precipitated by the alkaloidal reagent.
Protein Solution (2 ml) in a test tube ↓ Add Sulfosalicyclic Acid (drop by drop) ↓ Shake & Stand for 5 min	A white precipitate is formed.	Protein is precipitated by the alkaloidal reagent.

Note:

- This test using sulfosalicylic acid is commonly employed for the preliminary screening of urine for the presence of proteins. It is also used to identify proteins in CSF.
- For the estimation of blood constituents photometrically, proteins interfere with the analysis. This is avoided by an initial protein precipitation by alkaloidal reagents.

Precipitation by heavy metal ions

Principle:

2 [protein chain with $C(=O)-\ddot{O}:^{\ominus}$ and $\ddot{N}H_2$] + Hg^{2+} → [Hg complex with H_2N and NH_2]

Precipitates

- Proteins exist as negatively charged ions in pH higher than their isoelectric pH
- To such a solution if salt of heavy metals is added, positively charged metal ions can complex with protein anion
- Metal proteinates are formed which get precipitated.

Test	Observation	Inference
Protein Solution (2 ml) in a test tube ↓ Add 10% Lead Acetate (drop by drop) ↓ Shake & Stand for 5 min	A white precipitate is formed.	Protein is precipitated by heavy metals such as lead.
Protein Solution (2 ml) in a test tube ↓ Add 5% mercuric nitrate solution (drop by drop) ↓ Shake & Stand for 5 min	A white precipitate is formed.	Protein is precipitated by heavy metals such as mercury.

Precipitation by heat and acid (heat and acetic acid test)

Principle: On heating, the protein loses its structure and becomes denatured to form a coagulum. It is precipitated after the addition of acetic acid, which provides the suitable pH to get the maximum precipitate.

Test	Observation	Inference
Take the solution to be tested up to 3/4th of the test tube ↓ Hold the tube over a flame in a slanting position ↓ Boil the upper portion of the test tube ↓ Lower half serves as control ↓ Add 1% acetic acid drop by drop	A white coagulum is seen at the upper heated portion.	Indicates the presence of heat coagulable protein like albumin.

Note:

- To provide a suitable pH to get maximum precipitate.
- To differentiate between protein and interfering substances, mainly phosphates.
- If the precipitate persists and deepens after the addition of acetic acid, it is due to proteins. If it disappears, it indicates the presence of phosphates.

Color Reactions for Proteins

Biuret test

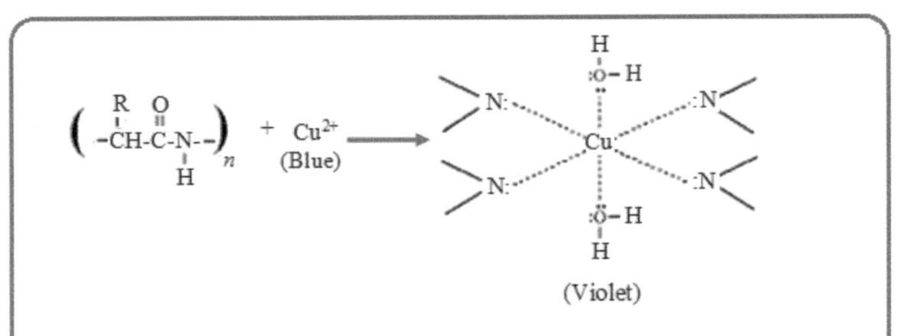

- Cupric ions in alkaline medium form a violet colored complex with peptide bond nitrogen.
- Copper sulfate is converted to cupric hydroxide which chelates with peptide linkage in proteins to give the purple color.

Test	Observation	Inference
Protein solution (2 ml) in a test tube. ↓ Add 2 ml of 5% sodium hydroxide. ↓ Mix and add of 1% copper sulfate (two drops) ↓ Shake & Stand for 5 min	Violet color is formed.	Presence of peptide linkage.

Note:

- Insoluble proteins such as keratin give negative Biuret test.
- Care must be taken that not more than two drops of dilute copper sulfate be added; otherwise, blue color (due to the excess formation of cupric hydroxide) will develop instead of violet color.
- Magnesium sulfate and ammonium sulfate interfere with this test.

Application:

- It is a common and delicate test for identifying proteins in a biological material.

Ninhydrin test

Principle:

Ninhydrin + Amino acid → Aldehyde + NH_3 Hydrindantin + CO_2

Hydrindantin + NH_3 → Bluish purple colored complex

- Ninhydrin is a powerful oxidizing agent and causes oxidative decarboxylation of α amino acids producing an aldehyde.
- The reduced Ninhydrin (hydrindantin) then reacts with ammonia and another molecule of Ninhydrin and produces bluish purple colored complex.

Test	Observation	Inference
Protein solution (2 ml) in a test tube. ↓ Add 10% drops of ninhydrin ↓ Heat to boiling ↓ Shake & Stand for 5 min.	Bluish purple color is formed.	Alpha amino acid is present.

Note:

- This test is positive for all α-amino acids.
- Amino acids which do not contain α-amino group, such as proline and hydroxy proline, give a yellow color with ninhydrin.
- Glutamine and asparagine give brown color with ninhydrin.
- This test is used as an additional test to confirm the presence of protein in a solution.
- This test is positive for all amino acids containing free amino and carboxylic groups. Hence, it is positive for proteins, peptones, and peptides.
- If the Biuret test is negative and the ninhydrin test is positive in a given solution, it indicates that free α-amino acids are present in the given solution.

Xanthoproteic test

Principle:

$$HO-C_6H_4-CH_2-CH(NH_2)-C(=O)-OH + HNO_3 \longrightarrow HO-C_6H_3(O_2N)-CH_2-CH(NH_2)-C(=O)-OH$$

(Yellow Colored Complex)

- The benzene ring system in tyrosine and tryptophan undergo nitration on treatment with strong nitric acid at elevated temperature forming a yellow precipitate.
- The yellow precipitate turns orange due to ionization, in alkaline medium.

Test	Observation	Inference
Protein solution (3 ml) in a test tube ↓ Add conc nitric acid (1 ml) ↓ Mix. & Heat the solution (1 min) ↓ Cool under tap water ↓ 40 % NaOH (2 ml) ↓ Observe the color	In the acid medium, yellow color is formed.  In the alkaline medium, orange color is formed. 	Tyrosine and tryptophan aromatic amino acids are present. Indicates the absence of tyrosine and tryptophan aromatic amino acids

Note:

- This is a specific test for aromatic amino acids.
- The yellow color is due to the formation of the nitro derivatives of benzene-ring-containing amino acids.

Millon's test (Cole's mercuric nitrite test)

Principle:

Tyrosine

Red color precipitate

- Sodium nitrite reacts with Sulfuric acid to form nitrous acid (reacting acid).
- The protein gets precipitated by the mercuric sulfate.
- The reacting groups (phenol group of tyrosine) which get exposed on boiling, reacts with nitrous acid to form mercury phenolate. This gives red color precipitate

Millon's reagent: 10% mercuric sulfate (prepared in 10% Sulfuric acid) and 1% sodium nitrite.

Test	Observation	Inference
Take of protein solution (1 ml) in a test tube ↓ Add 10% mercuric sulfate ($HgSO_4$) (1 ml) ↓ Boil gently cool it under tap water ↓ Add 1% sodium nitrate ($NaNO_3$) solution ↓ Mix and observe	Red coagulum is formed.	Indicates the presence of tyrosine (hydroxy phenyl group)

Note:

- This test is specific for tyrosine (hydroxy phenyl group).
- This test cannot be employed to detect tyrosine in urine because chlorides, which are normally present in urine, interfere with the reaction by forming unionized mercuric chloride.
- This test is given by phenols or phenolic substances such as salicylic acid.

Hopkins–Cole test (aldehyde test)

Principle:

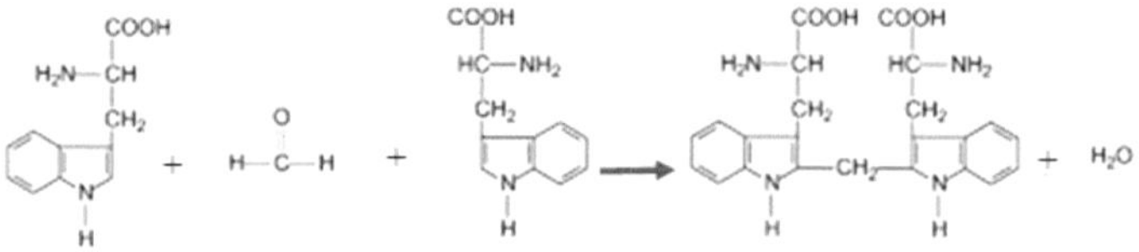

Tryptophan + Formaldehyde + Tryptophan → Colorful Condensation Product + H_2O

- This test is specific for Tryptophan
- Mercuric sulfate causes mild oxidation of indole group of tryptophan, which condenses with an aldehyde to give the colored complex.

Test	Observation	Inference
Take Protein solution (3 ml) in a test tube ↓ Add 0.2% formalin (2 drops) ↓ 10% mercuric sulfate ($HgSO_4$) (2 drop) ↓ Add Sulfuric Acid (3 ml) along the side of test tube	Violet and purple colors are formed at the junction of the two liquids. 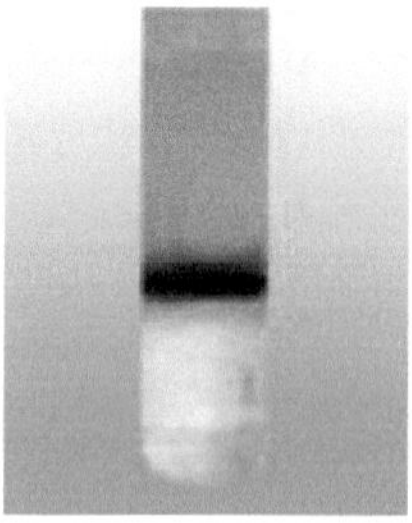	Indicates the presence of tryptophan (indole ring)

Note:

- It is a specific test for indole nucleus/ring.
- Sulfuric acid with mercuric sulfate is used as an oxidizing agent in this reaction.
- Tryptophan is an essential amino acid, and its presence indicates a good nutritive value of the protein.
- Casein and egg albumin give a positive test.

Sakaguchi's test (guanidine group)

Principle:

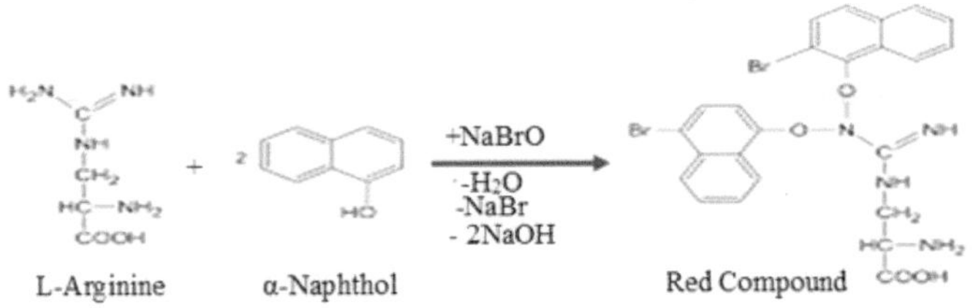

L-Arginine + 2 α-Naphthol $\xrightarrow[-H_2O,\ -NaBr,\ -2NaOH]{+NaBrO}$ Red Compound

- In alkaline medium α- naphthol combines with the guanidine group of arginine to form a complex which is oxidized by sodium hypobromite to form a bright red colored complex.

Test	Observation	Inference
Take protein solution (3 ml) in a test tube ↓ Add 40% NaOH (5-6 drops) ↓ Molish reagent (4 drop) ↓ mix and add Bromine water (10 drops) ↓ Observe the color	Bright red color is formed. 	Indicates the presence of arginine

Note:

- The test is specific for the guanidine group of arginine.
- Sodium hypochlorite can be used instead of sodium hypobromite.
- Avoid the addition of excess of α-naphthol, as it masks the development of color.

Sulfur test for cystine and cysteine

Principle:

$$R\text{-}SH + 2\,NaOH \longrightarrow R\text{-}OH + Na_2S + H_2O$$

$$Na_2S + (CH_3COO)_2Pb \longrightarrow PbS + 2CH_3{+}COONa$$

- On boiling with sodium hydroxide, the sulfur present in the protein is converted to inorganic sodium sulfide
- This reacts with lead acetate to form a black precipitate of lead sulfide

Test	Observation	Inference
Take Protein solution (2 ml) in a test tube ↓ Add 40% NaOH (1 ml) ↓ Boil for one minute ↓ Cool it under tap water ↓ Then and 5 drops of lead acetate	Black and brown color precipitate is formed.	Indicates the presence of cystine and cysteine residues

Note:

- This test is specific for the –SH (thiol) group of cysteine and cystine.
- It is a test for sulfur-containing amino acids.

Pauly's test (histidine and tyrosine)

Principle:

$H_2N-C_6H_4-SO_3H + HNO_2 + HCl \longrightarrow Cl^-N_2^+-C_6H_4-SO_3H + 2H_2O$

Sulphanilic acid → Diazonium salt

$Cl^-N_2^+-C_6H_4-SO_3H + C_6H_4(OH)CH_2CH(NH_2)COOH \longrightarrow (HO)(CH_2CH(NH_2)COOH)C_6H_3-N{=}N-C_6H_4-SO_3H + HCl$

Diazonium salt + Tyrosine → Azo dye

- This test is specific for the detection of tyrosine or histidine
- The reagent used for this test contains sulphanilic acid dissolved in hydrochloric acid
- Sulphanilic acid upon diazotization in the presence of sodium nitrite and hydrochloric acid results in the formation of a diazonium salt
- The diazonium salt give a red-colored chromogen (azo dye)

Test	Observation	Inference
Take 0.5% sulfanilic acid (1 ml) in test tube ↓ Add 0.5% sodium nitrate solution (1 ml) ↓ Leave in the cold for 3 min. ↓ Make the solution alkaline by sodium carbonate solution (2 ml) ↓ Observe whether red-colored products are formed or not.	Precipitate of cherry red color is formed.	Indicates the presence of histidine

Note:

- It is also positive for the phenolic hydroxyl group.
- A positive Pauly's test and a negative Millon's test indicate the presence of histidine.

Test for organic phosphorus (Neumann's test) (test with casein solution)

Principle: On boiling with strong sodium hydroxide, the organic phosphate present in phosphoproteins is released as inorganic phosphate. Inorganic phosphate reacts with ammonium molybdate in the presence of nitric acid (acidic media) to form a canary yellow precipitate of ammonium phosphomolybdate.

Test	Observation	Inference
Take protein solution (5 ml) in a test tube ↓ Add 40% sodium hydroxide (0.5 ml). ↓ Heat strongly ↓ Cool under tap water. ↓ Add conc. Nitric acid (0.5 ml). ↓ Filter to the filtrate ↓ Add pinch of solid ammonium molybdate ↓ Warm gently & observe the color	Canary yellow precipitate is formed.	Indicates the presence of phosphoprotein

Note:

- This test detects the presence of organic phosphorus in casein.
- Casein is a phosphoprotein.
- Ammonia is added to remove the sulfate ions.
- Nitric acid provides the acidic medium.

Table 7.1 Color reaction of specific amino acids

Test	Reagents	Results	Group Responsible	Importance
Biuret test	NaOH + $CuSO_4$	Violet	Peptide linkage	Tripeptides up to protein
Ninhydrin test	Triketohydrin-dene hydrate	Blue or purple	Free amino and free COOH	Amino acid, peptides in determining amino acids
Xantho-proteic test	Conc. HNO_3	Lemon yellow	Benzene ring	Tyrosine, phenylalanine, tryptophan
Millon's test	Hg in HNO_3	Red	Hydroxy-phenyl group	Tryptophan
Hopkins test	Glyoxylic acid and conc. H_2SO_4	Violet ring	Indole group	Tryptophan
Sakaguchi's test	10% NaOH and naphthol, alkaline hypobromite	Intense red color	Guanidine	Arginine
Sulfur test	KOH, $Pb(OAc)_2$	Black ppt	Sulfur	Cystine, cysteine, and methionine
Pauly's test	Diazotized sulfanilic acid	Red	Phenols and imidazole	Tyrosine, histidine

Multiple Choice Questions

1. Biuret reaction is specific for
   (a) $-CSNH_2$ group
   (b) $-(NH)NH_2$ group

   (c) –CONH linkages
   (d) All of these

Ans: c

2. Sakaguchi's reaction is specific for
   (a) Tyrosine
   (b) Proline
   (c) Cysteine
   (d) Arginine

Ans: d

3. Million-Nasse's reaction is specific for
   (a) Tryptophan
   (b) Phenylalanine
   (c) Tyrosine
   (d) Arginine

Ans: c

4. Ninhydrin with the evolution of CO_2 forms a blue complex with
   (a) α-Amino acids
   (b) Peptide bond
   (c) Serotonin
   (d) Histamine

Ans: a

5. Casein, the milk protein, is
   (a) Nucleoprotein
   (b) Chromoprotein
   (c) Glycoprotein
   (d) Phosphoprotein

Ans: d

6. The technique used for the purification of proteins that can be made specific for a given protein is
   (a) Affinity chromatography
   (b) Gel filtration chromatography
   (c) Ion exchange chromatography
   (d) Electrophoresis

Ans: a

7. Aromatic amino acids can be detected by
   (a) Sakaguchi's reaction
   (b) Millon–Nasse reaction
   (c) Hopkins–Cole reaction
   (d) Xanthoproteic reaction

Ans: d

8. A/G ratio is
   (a) Strength of proteins
   (b) Ratio of ceruloplasmin
   (c) Ratio of serum proteins
   (d) None of these

Ans: c

9. The useful reagent for the detection of amino acids is
   (a) Dichlorophenol indophenol
   (b) Ninhydrin
   (c) Molisch reagent
   (d) Biuret

Ans: d

10. The amino acid that gives yellow color with ninhydrin in paper chromatography is
    (a) Tyrosine
    (b) Proline
    (c) Alanine
    (d) Tryptophan

Ans: d

11. Millon's test is for the identification of
    (a) Tyrosine
    (b) Tryptophan
    (c) Proline
    (d) Arginine

Ans: b

12. The Hopkins–Cole test is for the identification of
    (a) Arginine
    (b) Tyrosine

(c) Tryptophan
(d) Cysteine

Ans: [illegible]

13. Which of the followings gives a positive test for ninhydrin?
    (a) Reducing sugars
    (b) Triglycerides
    (c) Alpha amino acids
    (d) Esterified fats

Ans: c

14. In the nitroprusside test, amino acid cysteine produces a
    (a) Red color
    (b) Blue color
    (c) Yellow color
    (d) Purple color

Ans: a

15. The pH of albumin is
    (a) 2.6
    (b) 4.7
    (c) 5.2
    (d) 6.1

Ans: b

16. The solubility of most proteins is lowered at high salt concentrations. This process is called
    (a) Salting in process
    (b) Salting out process
    (c) Isoelectric focusing
    (d) None of these

Ans: b

17. The xanthoproteic test is positive for proteins containing
    (a) Aromatic amino acids
    (b) Sulfur amino acids
    (c) α-Amino acids
    (d) Aliphatic amino acids

Ans: a

18. The indole group of tryptophan responses positively to
    (a) Glyoxylic acid
    (b) Schiff's reagent
    (c) Biuret test
    (d) Resorcinol test

Ans: a

19. The guanidine group of argentine gives positive test with
    (a) Lead acetate
    (b) Sakaguchi's reagent
    (c) Trichloroacetic acid
    (d) Molisch's reagent

Ans: b

20. Biuret test is specific for
    (a) Phenolic group
    (b) Two peptide linkages
    (c) Imidazole ring
    (d) None of these

Ans: b

21. The thiol group of cysteine gives red color with
    (a) Sodium acetate
    (b) Lead acetate
    (c) Barfoed's reagent
    (d) Sodium nitroprusside

Ans: d

Chapter 8

Lipid Analysis

Lipids are esters of fatty acids with alcohol, which are soluble in organic solvents and insoluble in water.

Simple Lipids

Simple lipids are made up of fatty acids and alcohol subdivided into two:

1. **Fat and oil:** The alcohol present is glycerol.
2. **Waxes:** A high molecular weight alcohol like acetyl alcohol is present.

Compound Lipids

Apart from fatty acids and alcohols, additional groups are also present in compound lipids. These lipids are of three types:

1. **Phospholipids:** These are made up of fatty acid alcohol, phosphoric acid, and a nitrogenous base. For example, lecithin, cephalin, sphingomyelin, and plasmalogen.
2. **Glycolipids:** Fatty acids, carbohydrates, sphingosine, and alcohol are present in glycolipids. They are subdivided into two types:

Clinical Biochemistry: A Laboratory Guide
Rooma Devi, Aman Chauhan, Simmi Kharb, and Chandra Shekhar Pundir

ISBN 978-981-4968-75-1 (Hardcover), 978-1-003-45566-0 (eBook)
www.jennystanford.com

a. **Cerebrosides:** They contain fatty acids, glucose/ galactose, and sphingosine.
b. **Gangliosides:** They contain fatty acids, sphingosine, N-acetyl numeric acid, and glucose/galactose.

3. **Lipoproteins:** They are lipids with protein as prosthetic groups, e.g., chylomicron, VLDL, HDL, LDL. They help in the transport of lipids in aqueous media.

Tests for Qualitative Analysis of Lipids

Solubility Test

Principle: This test is based on the solubility of lipids in organic solvents and insolubility in water.

Procedure	Observation	Inference
Take 4 test tubes ↓ 3 ml H_2O, Ether chloroform, Benzene CCl_4 in each test tube ↓ Add Oil into each test tube (5 drops) ↓ Shake well & allow to stand	 Oil and H_2O separate quickly in the first test tube, whereas a clear solution is formed in the case of organic solvents.	Lipids are insoluble in water (polar solvent) and soluble in organic solvents (nonpolar solvent).

Reagents: Ether, chloroform, benzene, CCl_4

Grease Spot Test/Translucent Spot Test

Principle: All lipids are greasy in nature, so this test is used as a group test for lipids. Lipids cannot wet a filter paper, unlike water. A spot of water disappears from a filter paper, whereas a spot of lipid remains there.

Procedure	Observation	Inference
Take 3 ml of Ether in test tube ↓ Add Oil (5 drops) ↓ Shake well ↓ Put Solution (1-2 drops) on Filter paper	Oil and H_2O separate quickly in the first test tube, whereas a clear solution is formed in the case of organic solvents.	Lipids are insoluble in water (polar solvent) and soluble in organic solvents (nonpolar solvents).

Reagents: Ether, chloroform, benzene, CCl_4

Saponification Test

Principle: When oil and fat are boiled with alkali, both are hydrolyzed and fatty acids are liberated from salts with alkali, called soap. The process is known as saponification.

$$\text{Triacylglycerol} + 3\ \text{NaOH} \longrightarrow \text{Glycerol} + 3\ \text{R-COONa (soaps)}$$

Procedure	Observation	Inference
Take Oil (5 drops) + Alkali (5 ml) in test tube ↓ Boil the solution in water bath (5 min.) ↓ At last, add ethanol	A white precipitate of insoluble sodium soap is formed.	A sodium salt of fatty acid is formed.

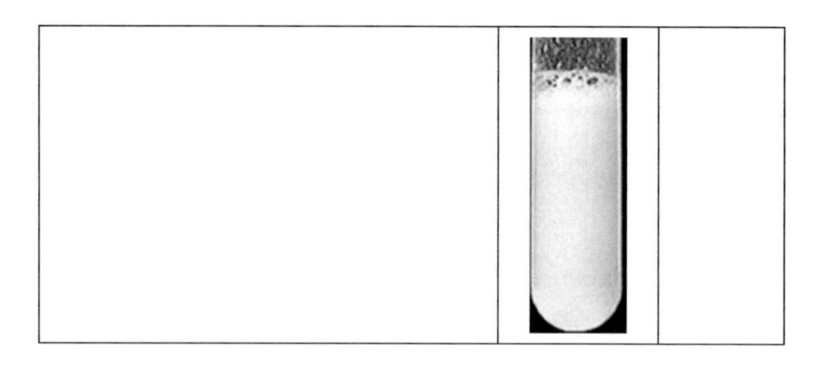

Emulsification Test

Principle: When oil and water are shaken together, the oil is broken down into small droplets, which are dispersed in the water. This is acknowledged as an oil-in-water emulsion. Water, due to its high surface tension, has a tendency to close together and separate as a layer. Bile salt and Na_2CO_3 soap solution decrease the surface tension. So, they are great emulsifying agents.

Emulsification is used in the analysis of lipids.

Procedure	Observation	Inference
Take Four (A, B, C, D) test tube ↓ Take 5 ml of each Water, Bile salt, Na_2CO_3, Soap solution in separate test tube ↓ Add Oil (5 drops) each test tube ↓ Shake well and allow it to stand for 10 min	Oil and H_2O separate quickly. Separation does not take place with soap, bile salt, and Na_2NO_3.	Emulsification is unstable in water (polar solvent) and stable with soap, bile salt, and Na_2NO_3.

Reagents: 0.5% Na_2NO_3 solution in water + 5% bile salt solution

Dustan's Test (for Detection of Glycerol)

Principle: Borax is hydrolyzed into NaOH (strong acid) and boric acid (weak acid) after the addition of glycerol. It reacts with boric acid and forms a strong acid (glyceroboric acid).

Procedure	Observation	Inference
3 ml of Boxer solution in the test tube ↓ Add drop of phenolphthalein indicator pink color procedure indicate medium is alkaline ↓ Add 20% glycerol drops by drops until pink color disappear indicate medium is acidic ↓ Heat reappear that indicate medium has become alkaline again	When the medium is alkaline, pink color disappears. When the medium is acidic, color appears.	Presence of glycerol

Reagents: Borax solution and phenolphthalein as indicator.

Acrolein Test (for Detection of Glycerol)

Principle: On heating with $KHSO_4$, glycerol is dehydrated to form air unsaturated aldehyde (acrolein).

Oil or Fat $\xrightarrow{\blacktriangle}$ Glycerol + Fatty acid

$$CH_2OH{-}CHOH{-}CH_2OH \xrightarrow[\blacktriangle]{KHSO_4} CH_2{=}CH{-}CHO + 2H_2O$$

Procedure	Observation	Inference
Take dry test tube ↓ Add glycerol (1-2 drops) ↓ Add a pinch of $KHSO_4$ ↓ Heat it	A pungent smell is produced.	Presence of glycerol

Reagent: Solid $KHSO_4$

Salkowski's Test (Cholesterol Analysis)

Principle: A sample is dissolved in chloroform and an equal quantity of H_2SO_4 is added. If cholesterol is present, the solution become blue red and change to violet pink and H_2SO_4 turn out to be red with green fluorescence.

Procedure	Observation	Inference
Take cholesterol (2 ml) in test tube ↓ Add conc. H_2SO_4 (2 ml) ↓ Shake well & Allow to stand (5 min.)	Two layers are separated. The chloroform layer is cherry red. The acid layer is fluorescent green.	Presence of cholesterol

Reagent: Concentrated H_2SO_4

Liebermann–Burchard Test

Principle: Cholesterol reacts with concentrated H_2SO_4 and acetic anhydride, leading to the formation of 2,4- or 3,5-cholestroldiene. H_2SO_4 and acetic anhydride act as a dehydrating and oxidizing agent.

Procedure	Observation	Inference
Take cholesterol (2 ml) in test tube ↓ Add acetic anhydride (10 drops) ↓ Add conc. H_2SO_4 (1-2 drops) ↓ Mix well	Red color develops, followed by blue, and finally the whole solution becomes green.	Presence of cholesterol

Reagents: Concentrated H_2SO_4 and acetic anhydride

Hübl's Iodine Test (Test for Unsaturation)

Principle: An unsaturated fatty acid absorbs iodine at double bonds till all bonds are saturated with iodine.

Procedure	Observation	Inference
CH_3Cl (3 ml) in test tube ↓ Add Hubli's Reagent (drop by drop) ↓ Mix well ↓ Allow to stand (30 sec)	The color of iodine disappears.	Shows degree of unsaturation

Reagents: Iodine (26 g), mercuric chloride (30 g), and C_2H_5OH (1000 ml)

Chapter 9

Lipid Profile

Some investigations routinely done in the laboratory for studying lipid profile are provided in Table 9.1.

Table 9.1 Routine investigation done in the laboratory for lipid profile

Name of Investigation	Method Name	Biological References
S. triglycerides	Enzymatic	60–160 mg/dl
S. cholesterol	CHOD-PAP	130–230 mg/dl
S. HDL cholesterol	Direct method	30–60 mg/dl
S. LDL cholesterol	By calculation	Up to 160 mg/dl
S. VLDL cholesterol	By calculation	16–32 mg/dl

Cholesterol

Cholesterol is one of the causative factors of heart diseases, thrombosis, and cerebral hemorrhage. Cholesterol is found in two forms:

- 30% free cholesterol
- 70% esterified cholesterol

Clinical Biochemistry: A Laboratory Guide
Rooma Devi, Aman Chauhan, Simmi Kharb, and Chandra Shekhar Pundir

ISBN 978-981-4968-75-1 (Hardcover), 978-1-003-45566-0 (eBook)
www.jennystanford.com

Lipogram

Total cholesterol	200–250 mg/dl
S. triglycerides	25–170 mg/dl
S. HDL cholesterol	35–60 mg/dl
S. LDL cholesterol	Up to 150 mg/dl
S. VLDL cholesterol	20–35 mg/dl

Estimation of Serum Cholesterol

- Cholesterol is the most abundant sterol in the human tissues and body fluids.
- It is transported bound to lipoproteins, mainly HDL, LDL, and VLDL.

Collection of Sample

- Patients should come fasting for 12 h. This is because dietary cholesterol tends to increase serum cholesterol.
- From the antecubital vein, 5 ml of blood is taken.
- The blood should be allowed to clot properly for at least 30 min before separating the serum since even minute hemolysis interferes with colorimetric analysis.
- Lipid levels are affected by behavioral factors such as recent diet, alcohol consumption, exercise, and medication (oral contraceptives, steroids, lipid-lowering drugs).

Cases When Cholesterol Testing Is Important

- Family history of excessive cholesterol/coronary heart disease
- Overweight/obese
- Drinking alcohol frequently
- Smoking cigarettes
- Leading an inactive lifestyle
- Diabetes, kidney disease, polycystic ovary syndrome, or an underactive thyroid gland

Other Methods of Estimation

1. Colorimetric Methods

(a) Uranyl acetate method

Principle: The cholesterol present in serum gets precipitated with uranyl acetate, ferric chloride, and glacial acetic acid (uranyl acetate reagent). This eliminates all the interfering chromogens, including bilirubin. The supernatant is made to react with concentrated sulfuric acid, which acts as a dehydrating agent, and it forms cholestene polyene carbonium ion. It is red in color. The intensity of color is directly proportional to the concentration of cholesterol in the given sample. The absorbance is measured at 560 nm.

(b) Zak Zlatkis method

Principle: Cholesterol reacts with ferric chloride, converting ferric ions to ferrous ions, the concentration of which is directly proportional to the concentration of cholesterol. This method can estimate free as well as esterified cholesterol. Digitoxin is used to extract free cholesterol.

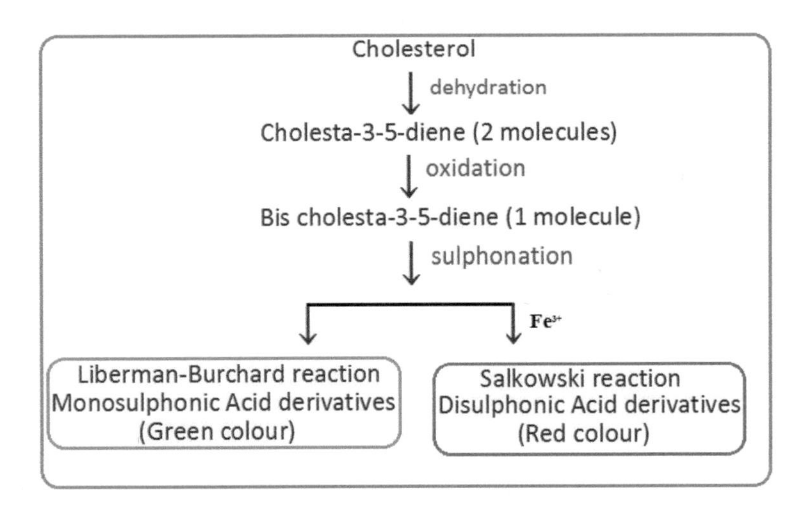

(c) Liebermann–Burchard reaction

Principle: On treating cholesterol with acetic anhydride and concentrated H_2SO_4 in anhydrous conditions, intense blue color develops due to the formation of polyunsaturated hydrocarbons. The addition of 2,5-dimethylbenzene sulfonic acid disperses proteins.

2. **Enzymatic Method (CHOD-PAP)**

Principle: The absorbance of chromophore quinoneimine is directly proportional to the concentration of cholesterol in the sample. The absorbance is measured at 505 nm.

$$\text{Cholesterol ester} + H_2O \xrightarrow{\text{Cholesterol esterase}} \text{Cholesterol} + \text{Free fatty acids}$$

$$\text{Cholesterol} + O_2 \xrightarrow{\text{Cholesterol oxidase}} \text{Cholestene-3-one} + H_2O_2$$

$$2\,H_2O_2 + \text{Phenol} + \text{4-aminoantipyrine} \xrightarrow{\text{Peroxidase}} \text{Quinoneimine} + H_2O_2$$

The normal range is 140–210 mg%.

Procedure:

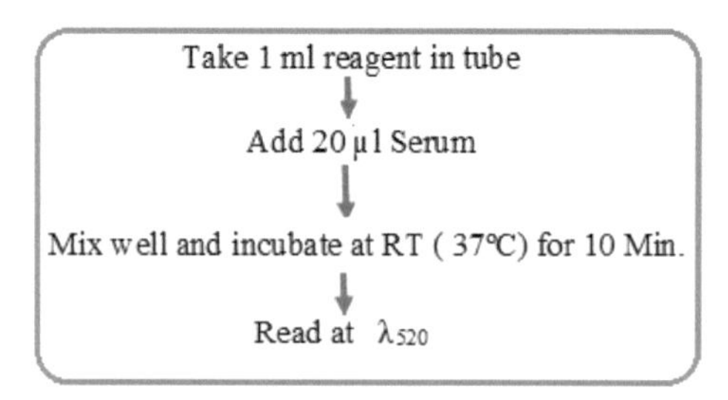

Pipette into labeled test tubes/cuvettes, as shown in Table 9.2.

Table 9.2 Labeled test tubes/cuvettes

	Labeled Test Tubes/Cuvettes		
Reagents	**Blank (B)**	**Test (T)**	**Standard (STD)**
Distilled H_2O	1000 µl	—	—
Sample supernatant	—	20 µl	—
Standard supernatant	—	—	20 µl
Cholesterol reagent R1	1000 µl	1000 µl	1000 µl

Calculation:

1. **Using a standard**

$$\frac{\text{Absorbance of test}}{\text{Absorbance of standard}} \times \text{Concentration of standard (mg/dl)}$$

2. **Using a factor**

Wavelength	mmol/l	mg/dl
Hg 546 nm	8.27	320
500 nm	5.43	210

Note: If cholesterol concentration is above 901 mg/dl, the sample should be diluted 1 + 2 with 0.9% NaCl. Multiply the result by 3.

- **LDL:** The serum level of LDL is calculated by calculation method (by Friedewald formula).
  LDL is calculated as follows:

$$\text{LDL} = \text{Total Cholesterol} - \frac{\text{Triglyceride}}{5} - \text{HDL (mg/dl)}$$

- **VLDL:** The serum levels of VLDL is also calculated by calculation method (by Friedewald formula).
  VLDL is calculated as follows:

$$\text{VLDL} = \frac{\text{Triglyceride}}{5} \text{(mg/dl)}$$

Advantages of enzymatic method (CHOD-PAP):

- It is more sensitive and specific for cholesterol.
- Its linearity is up to 700 mg%.
- It requires very low sample volume, i.e., 10 µl.
- It is less time consuming (10 min).
- It requires only a single, noncorrosive reagent.
- No deproteinization is needed, and no interference by icteric is observed.

Hypercholesterolemia: It is mainly seen in

- Primary hyperlipoproteinemia types I–IV
- Nephrotic syndrome
- Myxedema
- Obstructive jaundice
- Diabetes mellitus
- Xanthomatosis

In nephrotic syndrome when oedema is present, values up to 600–700 mg% are common and may occasionally reach and exceed 1000 mg%. In primary biliary cirrhosis, values may reach up to 1000–2000 mg%.

Hypocholesterolemia: It is seen in

- Pernicious anemia and other anemias
- Hemolytic jaundice
- Malabsorption syndrome
- Severe malnutrition
- Acute infections

Very low values may be seen in abetalipoproteinemia and familial hypobetalipoproteinemia. Therapeutic reduction is seen during the administration of lipid-lowering drugs such as clofibrate and cholestyramine and drugs such as nicotinic acid and D-thyroxine. Sometimes it may accompany chlorpropamide and phenformin therapy.

Ratio of Free and Esterified Cholesterol

Free cholesterol is normally 30% of the total cholesterol (range 20–40%). Although a change in the total cholesterol does not involve any change in the concentration of free cholesterol in diabetes mellitus, nephrotic syndrome, and myxedema, the percentage of free cholesterol rises in liver disease. In primary biliary cirrhosis, the increase is observed almost entirely in the free cholesterol and may be as high as 90%.

Triglyceride (TG)

Colorimetric Detection of Triglycerides by Trinder Method

Principle:

Triglycerides + H_2O $\xrightarrow{\text{Lipase}}$ Glycerol + FFA

Glycerol + ATP $\xrightarrow{\text{Glycerol Kinase}}$ Glycerol 3-PO_4 + ADP

Glycerol 3-PO_4 + O_2 $\xrightarrow{\text{Oxidase}}$ DHAP + H_2O_2

H_2O_2 + 4-aminoantipyrine + Phenol $\xrightarrow{\text{Peroxidase}}$ Quinonimine Dye (Red/Pink Color)

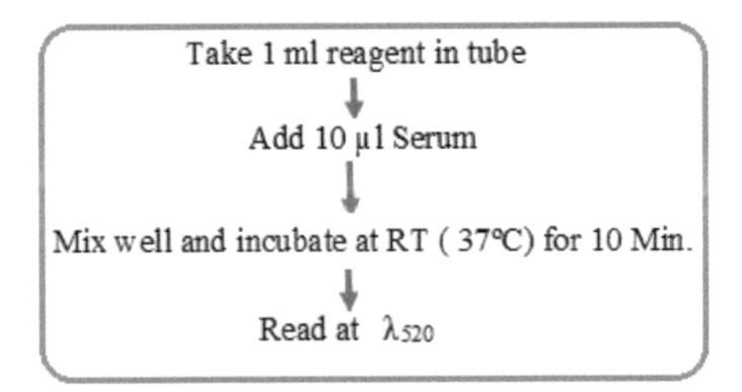

Preparation of working reagent: Reconstitute one vial of R1b with 15 ml of buffer R1a. Keep it stable for 21 days at +2 to +8°C. Set the wavelength of the spectrophotometer at 520 nm. Pipette into labeled test tubes/cuvettes as shown in Table 9.3.

Table 9.3 Procedure for estimation of serum TAG

	Labeled Test Tubes/Cuvettes		
Reagents	**Blank (B)**	**Test (T)**	**Standard (STD)**
Sample	—	10 µl	—
Standard	—	—	10 µl
Trigs R1b	1000 µl	1000 µl	1000 µl

Calculation:

1. **Using a standard**

$$\frac{\text{Absorbance of Test}}{\text{Absorbance of Standard}} \times \text{Concentration of Standard mg/dl}$$

2. **Using a factor**

Wavelength	mmol/l	mg/dl
Hg 546 nm	11.9	1048
520 nm	8.47	743

Note: If cholesterol concentration is above 1000 mg/dl, the sample should be diluted 1 + 4 with 0.9% NaCl. Multiply the result by 5.

HDL Cholesterol

Estimation of HDL is done by phosphotungstic acid method.

Principle: First precipitate LDL, VLDL, and chylomicrons by polyanions (Phosphotungstate) in the presence of metal ions like magnesium ions to leave HDL in the solution. Now centrifuge it and supernatant containing HDL will be used for HDL estimation.

Preparation of R1: Pre-dilute the precipitating reagent in the ratio 4:1 with redistilled water (dilute the contents of an 80 ml bottle with 20 ml redistilled water). Keep it stable up to the expiry date, particularly when stored at 15°C to 25°C.

Step 1: Precipitation

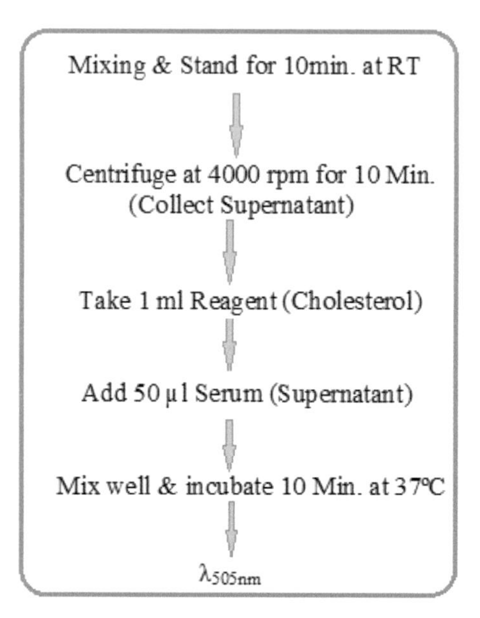

Pipette into labeled test tubes as shown in Table 9.4.

Table 9.4 Procedure for estimation of serum HDL

	Labeled Test Tubes	
Reagents	**Test (T)**	**Standard (STD)**
Sample	200 µl	—
Cholesterol standard	—	200 µl
Diluted precipitant R1	500 µl	500 µl

Table 9.5 Procedure for detection of HDL calorimetrically

	Labeled Test Tubes/Cuvettes		
Reagents	**Blank (B)**	**Test (T)**	**Standard (STD)**
Distilled H_2O	50 µl	—	—
Sample	—	50 µl	—
Standard	—	—	50 µl
Cholesterol R1	1000 µl	1000 µl	1000 µl

Note:

- Only clear supernatants generated after centrifugation should only used for estimation of total cholesterol, TAG, and HDL (not turbid supernatants).
- If turbid supernatants are caused by elevated triglyceride concentrations, the sample must be diluted 1 + 1 with 0.9% NaCl solution and the precipitation step should be repeated.
- Then the result has to be multiplied by 2.

Chapter 10

Liver Function Test

The liver is a critical organ that carries out many functions. It is referred to as the metabolic factory of the body. The liver function test routinely assesses the functioning of liver.

Table 10.1 Routine investigations done to assess liver functions

Name of Investigation	Method Name	Biological References
S. AST/ SGOT	UV kinetic	Up to 40 U/L
S. ALT/SGPT	UV kinetic	Up to 40 U/L
S. alkaline phosphate	PNP-AMP kinetic	39–117 U/L
S. protein	Biuret	6–8 g/dl
Albumin	BCG	3.8–4.4 g/dl
A/G ratio	By calculation	1–2
S. bilirubin total	Evelyn–Malloy	0.2–0.8 mg/dl
S. bilirubin direct	Evelyn–Malloy	0–0.2 mg/dl
S. bilirubin indirect	Evelyn–Malloy	0.2–0.7 mg/dl

Bilirubin

Bilirubin is the breakdown product of hemoglobin present in RBCs, from its heme part. It is produced in the reticuloendothelial system

Clinical Biochemistry: A Laboratory Guide
Rooma Devi, Aman Chauhan, Simmi Kharb, and Chandra Shekhar Pundir

ISBN 978-981-4968-75-1 (Hardcover), 978-1-003-45566-0 (eBook)
www.jennystanford.com

present in the spleen and liver. This bilirubin is unconjugated and insoluble in water. Unconjugated bilirubin gets bind to albumin and then it is taken up by liver through carrier-mediated process. Inside hepatocytes, it is bound by intracellular proteins, e.g., ligandin and Z protein. The enzyme UDP glucuronosyltransferase conjugates bilirubin, forming bilirubin di- and monoglucuronides. These forms are water soluble and filterable at the glomerulus.

The secretion of conjugated bilirubin in bile throughout the biliary canalicular membrane is rate limiting and is sensitive to liver damage. Bile is saved in the gall bladder from where it passes into the duodenum through the cystic and common bile duct. In the intestines, it is decreased by the bacterial action to urobilinogen. It is reabsorbed from the large gut, and most of it is excreted in bile through the liver via extrahepatic circulation.

Principle:

$NaNO_2 + HCl$ → $NaCl + HNO_2$

$HO_3S\text{-}C_6H_4\text{-}NH_2 + H_3O^+ \longrightarrow HO_3S\text{-}C_6H_4\text{-}\overset{+}{N}H_3 + H_2O$

Sulphanalic acid

$HO_3S\text{-}C_6H_4\text{-}\overset{+}{N}H_3 + HNO_2 \xrightarrow{+2H_2O} HO_3S\text{-}C_6H_4\text{-}\overset{+}{N}{\equiv}N \longleftrightarrow HO_3S\text{-}C_6H_4\text{-}N{=}\overset{+}{N}$

Nitrous acid

Dizonium ion

Conjugated bilirubin + $HO_3S\text{-}C_6H_4\text{-}N{=}\overset{+}{N}$ (Diazotised sulphanilic acid) ⟷ 2 Azobilirubin molecules

Total bilirubin: Bilirubin + Diazotized sulfanilic acid ⟶ Azobilirubin (purple or reddish pink)

Bilirubin direct: Directly react in an acidic medium (water soluble)

Bilirubin indirect: Indirect bilirubin solubilized after addition of surfactant

In the direct reaction, OD is measured within 1 min and is shown by conjugated bilirubin, as the reaction is very fast. In the indirect reaction, methyl alcohol is added, which solubilizes unconjugated bilirubin. It measures both conjugated and unconjugated forms.

Value of unconjugated bilirubin = Total (indirect) – Conjugated (direct)

Reagents:

Total Bilirubin Reagent (R_1)	
Sulphanilic acid	5 m mole/L
HCl	100 m mole/L
Surfactant	1%
Direct Bilirubin Reagent (R_2)	
Sulphanilic acid	10 m mole/L
HCl	100 m mole/L
Reagent (R_3)	
Sodium nitrate	144 m mole/l

Reagent preparation:

	Volume of working	Add		
Test	**Reagent**	R_1	R_2	R_3
Bilirubin Total	10 ml	10 ml	—	0.2 ml
Bilirubin Direct	10 ml	—	10 ml	0.1 ml
Keep the reagent 3 vial plugged after use				

Standard:

- An artificial standard is used i.e. methyl red.
- Methyl red is an azo dye with an absorbance curve very similar to azo bilirubin.
- Original standard is not used because it is photosensitive and thus unstable.

Observation:

Total bilirubin/direct bilirubin

Pipette into test tube	Blank	Standard	Test
Working reagent	500 µl	500 µl	500 µl
DW	25 µl	—	—

Pipette into test tube	Blank	Standard	Test
Standard	—	25 μl	—
Test	—	—	25 μl

- Mix well and incubate for 5 min at 37°C for total bilirubin/direct bilirubin. Read at 540–630 nm.

Calculations:

$$\frac{\text{Absorbance of Test}}{\text{Absorbance of Standard}} \times \text{Concentration of Standard}$$

Result:

- **Normal bilirubin level =** 0.2–0.8 mg%
- **Conjugated bilirubin =** 0.1–0.4 mg%
- **Unconjugated bilirubin =** 0.2–0.7 mg%

Other methods of estimation

- Jendrassik–Grof method
- Direct bilirubinometer
- Method using reflectance spectrophotometry
- HPLC method

Jaundice: It is the yellowish discoloration of the skin and mucosa due to hyperbilirubinemia. When the level is greater than 2 mg%, clinical jaundice, i.e., icterus is visible. If the level is between 0.8 and 2.0 mg%, subclinical/latent jaundice is observed.

Physiological Jaundice of Newborn: Neonatal jaundice occurs after birth within 24–48 h and clears in 4–5 days. It happens because of the relative lack of a conjugation system in newborn and the extra load of bilirubin due to excessive hemolysis.

Causes of Jaundice

1. **Pre-hepatic Jaundice**
   (a) In effective erythropoiesis, e.g., pernicious anemia
   (b) Increased hemolysis:
   - Incompatible blood transfusion

- Drug intake, e.g., sulfonamides, primaquin in patients suffering from glucose-6-phosphate dehydrogenase deficiency
- Abnormal RBCs, e.g., in spherocytosis, hemoglobinopathies, etc.

2. **Hepatic Jaundice**
   - Acute hepatitis
   - Chronic hepatitis
   - Cirrhosis of liver
   - Toxic drugs such as chloroform
   - Inborn error:
     - Gilbert's syndrome
     - Crigler–Najjar syndrome
3. **Post-hepatic Jaundice**
   - Stricture/stone/some growth in the bile duct
   - Carcinoma head of pancreas pressing the bile duct
   - Dubin–Johnson syndrome
   - Rotor syndrome

Table 10.2 Biochemical parameter in different types of jaundice

	Pre-hepatic	Hepatic	Post-hepatic
S. bilirubin	↑	↑	↑
Unconjugated bilirubin	↑↑↑	↑	↑
Conjugated bilirubin	↑	↑	↑↑↑
AST/ALT	↑	↑↑↑	↑
Alkaline phosphatase	↑	↑	↑↑↑
Urine urobilinogen	+	+	–
Bile salts	–	+/–	++
Stool color	Dark colored	Normal	Clay colored

Estimation of Total Protein

Blood contains a larger number of proteins that perform so many functions. Most enzymes, hormones, and clotting factor are proteins in nature. More than a hundred different proteins have been identified so far, and perhaps many more are yet to be identified. Principle serum proteins are albumins and globulins.

Methods of Estimation

1. Biuret method
2. Lowry method
3. Dye binding method
4. Turbidimetric method
5. Nephelometric method
6. Kjeldahl method

Biuret method: It is the most commonly used method in clinical laboratories for the estimation of total protein.

Principle:

Urea → (180°C) Biuret + NH_3 → (Cu^{2+}) Cu Complex (Violet Color)

- Compounds with two or more $-CO.NH_2$ groups linked together directly or through a carbon or nitrogen atom gives the reaction.
- Such compounds react with Cu^{++} ions in alkaline medium to form a purple-violet colored complex whose absorbance is read at 540 nm.
- The intensity of the color produced is proportional to the amount of protein undergoing the reaction.

Reagents: Biuret reagent, besides having copper sulphate and sodium hydroxide, contains sodium potassium tartrate to lex cupric ions and maintain their stability in alkaline medium; and iodine which acts as antioxidant.

Procedure:

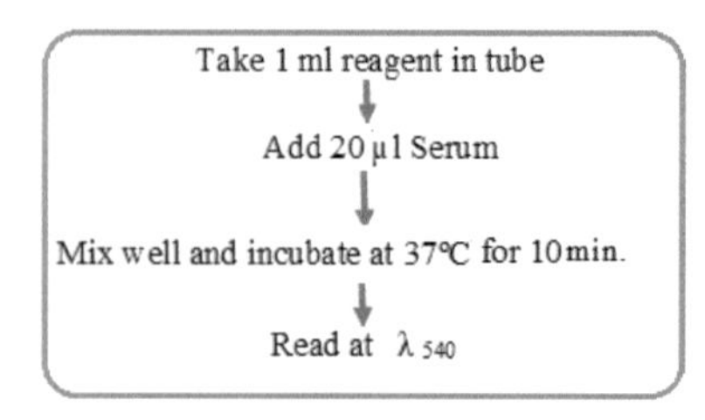

Observation:
Pipette into labeled test tubes/cuvettes as shown in Table 10.3.

Table 10.3 Procedure for the detection of serum protein

	Labeled Test Tubes/Cuvettes		
Reagents	**Blank (B)**	**Standard (S)**	**Test (T)**
Reagent	1 ml	1 ml	1 ml
Standard	—	20 μl	—
Serum	—	—	20 μl

Calculation:

$$\frac{\text{Absorbance of Test}}{\text{Absorbance of Standard}} \times \text{Concentration of Standard}$$

Result:

Normal Values in Plasma
Total proteins: 6.4–8.3 g/dl
Albumin: 3.5–5.2 g/dl
Globulins: 2.0–3.5 g/dl
A:G ratio: 1.5–2.5
Fibrinogen: 0.2–0.4 g/dl

- Most plasma proteins with the exception of immunoglobulins and protein hormones are synthesized in the liver.

Functions of Plasma Proteins

- Transport of fatty acids, bilirubin, uric acid, etc.
- Action as buffer
- Nutritive function
- To provide immunity to the body
- Participation in blood coagulation
- Maintenance of oncotic pressure
- Enzymatic action

Clinical Significance

Hypoproteinemia or decreased serum protein level (below normal range) usually happens due to reduced albumin levels and hence accompanied by reduced A:G ratio.

Causes:

Decreased input:

- Chronic malnutrition due to a protein-deficient diet over a long period
- Malabsorption, e.g., celiac disease and protein losing enteropathy

Decreased synthesis/increased catabolism:

- Chronic liver disorders leading to reduced synthesis of proteins
- Chronic diseases and high fever leading to increased protein catabolism

Increased output:

- Renal disorders, e.g., nephrotic syndrome leading to heavy proteinuria

False hypoproteinemia

Hemodilution happens due to water intoxication, salt retention, and massive I/V infusion.

Hyperproteinemia is an increase in the total serum protein above the normal levels. It occurs mainly due to a rise in plasma globulins and thus a reduced A:G ratio. It also happens due to multiple myeloma in which there is increased synthesis of Bence–Jones myeloma proteins (monoclonal gamma globulins).

Causes:

- Chronic infections, e.g., Kala azar.
- Multiple myeloma
- Collagen disorders
- Macroglobulinemia
- Hemoconcentration can occur due to either inadequate water intake or excessive water loss due to vomiting, diarrhea, and burns.

Disadvantages:

- The main disadvantage is its lack of sensitivity. The Biuret method cannot be used to estimate proteins less than 1 mg/ml, so it cannot be used to estimate CSF proteins.

- False high results may be obtained as small peptides also give this test positive.

Precautions: As the amount of serum to be added is very small, it should be added using low-volume pipettes of capacity 0.1 or 0.2 ml.

Other methods:

1. **Kjeldahl method:** In this method, the total nitrogen content of the sample is determined. A correction for non-protein nitrogen (NPN) is done by using a protein-free filtrate, whose value is deducted from that of total nitrogen. This method is rarely used for routine estimation nowadays as it is time consuming and requires special instrumentation.
2. **Folin–Ciocalteu (Lowry) method:** The Lowry method is based on the presence of tyrosine or tryptophan content of the protein. This method is hundred times more sensitive than the Biuret method, so it can be used for the estimation of proteins in biological fluids, e.g., CSF, as proteins are present in milligrams only.
3. **Turbidimetric/nephelometric method:** This is a quite sensitive method, which can be used to estimate CSF proteins.
4. **Dye binding methods:** These methods are based on the property of proteins to bind certain dyes, such as amido black IOB, Coomassie brilliant blue, and bromocresol green. They are mainly used for staining protein bands after electrophoresis and estimating serum albumin. They are pH-sensitive methods and are used to estimate the A:G ratio.

Albumin

Albumin is a major plasma protein that performs so many functions such as transport, maintenance of colloid osmotic pressure, buffering action, and nutrition. Thus, albumin acts as a complete protein.

Bromocresol green method (BCG method): Bromocresol form green blue complex when treated with albumin under acidic condition, which is then measured calorimetrically.

Principle:

Albumin + Bromocresol green —Acidic pH→ Green blue complex

Procedure:

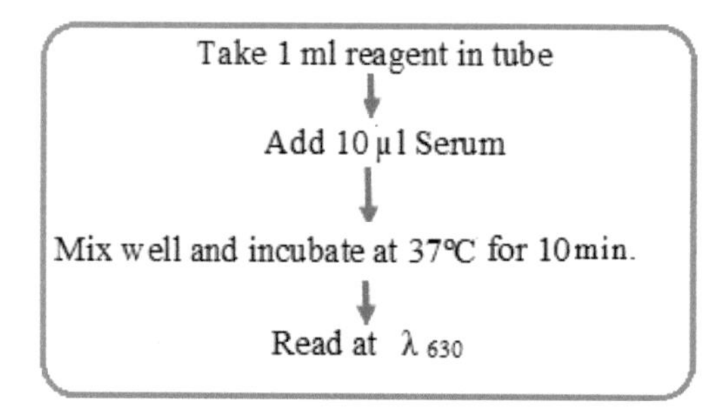

Observation:

Pipette into labeled test tubes/cuvettes as shown in Table 10.4.

Table 10.4 Procedure for estimation of serum albumin

Reagents	Labeled Test Tubes/Cuvettes		
	Blank (B)	**Standard (S)**	**Test (T)**
Reagent	1 ml	1 ml	1 ml
Standard	—	10 µl	—
Serum	—	—	10 µl

Calculations:

$$\text{Serum albumin} = \frac{A_{\text{Test}} - A_{\text{Blank}}}{A_{\text{Standard}} - A_{\text{Blank}}} \times \text{Conc. of Standard}$$

$$\text{Serum globulin} = \text{Serum total protein} - \text{Serum albumin}$$

Results:

Normal values in plasma:

Total proteins: 6.4–8.3 g/dl

Albumin: 3.5–5.2 g/dl
Globulins: 2.0–3.5 g/dl
A:G ratio: 1.5–2.5

Clinical Significance

Hypoalbuminemia:

- Liver disease
- Prolong starvation
- Malnutrition

Hyperalbuminemia:

- Shock
- Dehydration
- Hemoconcentration

Analbuminemia: It refers to the near total absence of plasma albumin in the homozygous state.

Microalbuminuria: It refers to an excretion of 30–300 mg of albumin/day. It is an early marker of hypertension and glomerular diseases (diabetic neuropathy).

Globulins

Globulin proteins include immunoglobins (antibodies) and many other proteins such as ceruloplasmin, transferrin, α-antitrypsin, and enzyme inhibitors.

Globulin Fractions in Serum

α_1-globulins: α_1-antitrypsin, etc.

α_2-globulins: ceruloplasmin

β-globulins: transferrin

γ-globulins: immunoglobulins IgG, IgA

Calculation:

$$\boxed{\text{Serum globulin} = \text{Serum total protein} - \text{Serum albumin}}$$

A:G ratio: It is calculated from the known albumin and globulin concentrations.

AST/SGOT

Aspartate transaminase (AST), also called serum glutamate oxaloacetate (SGOT), is widely distributed in the liver, heart, muscles, kidneys, brain, lungs, WBCs, etc. SGOT is present in highest concentration in liver.

Method: SGOT/AST is estimated by the kinetic UV IFCC (International Federation of Clinical Chemistry) standard.

Principle:

L-Aspartate+ 2Oxoglutrate $\xrightarrow{AST}$ Oxaloacetate + L-glutamate

Oxaloacetate + NADH $\xrightarrow{MDH}$ L-Lactate + NAD

Sample pyruvate + NADH $\xrightarrow{LDH}$ Lactate + NAD

AST- Aspartate aminotransaminase
MDH-Malate dehydrogenase
LDH- Lactate dehydrogenase

Procedure: Pipette into test tubes as shown below:

Reagents	**Volumes**
Working reagent	1000 μl
Test	100 μl

Mix all tubes, and incubate for exactly 1 min at 37°C.
Lagging time: 60 s
Reading time: 60 s
Read at λ_{max} = 340.

Calculation:

$$IU/L = \frac{(\Delta A/min.) \times TV \times 10^3}{V \times Absorption \times P}$$

P- Cuvette light path= 1cm
TV- Total reaction volume in µl
V- Sample volume in µl

Results:

Clinical Significance

An increase in the level of SGOT/AST causes the following:

- Heart diseases
- Myocardial infraction
- Liver disease

ALT/SGPT

Alanine amino transaminase (ALT), also called serum glutamate pyruvate (SGPT), is widely distributed in the liver, heart, kidneys, muscles, etc. SGPT is more specific to liver diseases.

Method: SGPT/ALT is estimated by the kinetic UV IFCC standard.

Principle:

$$\text{L-Alanine} + 2\ \text{Oxoglutrate} \xrightarrow{ALT} \text{Pyruvate} + \text{L-Lactate}$$

$$\text{Pyruvate} + \text{NADH} \xrightarrow{LDH} \text{L-Lactate} + \text{NADH}$$

ALT – Alanine aminotransaminase
LDH- Lactate dehydrogenase

Procedure: Pipette into test tubes as shown below:

Reagents	Volumes
Working reagent	1000 μl
Test	100 μl

Mix all tubes, and incubate for exactly 1 min at 37°C.
Lagging time: 60 s
Reading time: 60 s
Read at $\lambda_{max} = 340$.

Calculation:

$$IU/L = \frac{(\Delta A/min.) \times TV \times 10^3}{V \times Absorption \times P}$$

P- Cuvette light path= 1cm
TV- Total reaction volume in μl
V- Sample volume in μl

Results:

Clinical Significance

An increase in the level of SGPT/ALT leads to the following:

- Liver diseases
- Viral hepatitis
- Toxic hepatitis
- Liver cirrhosis
- Dengue

Alkaline Phosphate

Alkaline phosphate (ALP) is produced by the osteoblast of bones and is associated with the calcification of bones. ALP has different types of isoenzymes, which are found in the liver, bones, kidneys, intestines, and placenta.

Method: It is calorimetrically estimated by the King and King method.

$$\text{Phenyl Phosphate} \xrightarrow{\text{ALP}} \text{Phenol+ Pi}$$

$$\text{Phenol+ 4- Amino antipyrine} \xrightarrow{\text{Potassium ferrocyanide}} \text{Orange/Red Color}$$

Procedure:

Reagents	Blank	Standard	Control	Test
Reagent R1	500 μl	500 μl	500 μl	500 μl
DW	1500 μl	1500 μl	1500 μl	1500 μl
Mix well and incubate at 37°C for 15 min.				
Reagent R2	1000 μl	1000 μl	1000 μl	1000 μl
Serum	—	50 μl	—	—
Mix well and incubate at 37°C for 3 min.				
Reagent R3	—	50 μl	—	—
Serum	—	—	—	50 μl

Calculation:

$$= \frac{\text{OD of Test} - \text{OD of Control}}{\text{OD of Standard} - \text{OD of Bank}} \times \text{Concentration of Standard}$$

Results:

The normal range of ALP in blood is 29–130 U/L. An increased level is found in children.

Clinical Significance

An increase in the level of ALP leads to the following:

- Bone diseases
- Obstructive jaundice
- Bone carcinoma
- Paget disease

- Liver cirrhosis
- Disease of the intestinal tract
- Complications in the third trimester of pregnancy
- Diseases in growing children

A decrease in the level of ALP leads to the following:

- Severe anemia
- Malnutrition

Case Studies

Case Study 1

A 50-year-old man with a history of chronic alcoholism came to the hospital with complaints of protuberant abdomen (ascites) and edema in the feet. He also has a history of bleeding. The investigation report of the patient revealed decreased serum albumin levels, with 30 s of prothrombin time. Proteins were absent in the urine.

1. Which biochemical investigations will you do in the serum of this patient?
2. What is your diagnosis based on lab findings and history?
3. Analyze the given urine sample and interpret.

Case Study 2

An 8-year-old boy was admitted to the hospital with generalized edema. His urine was frothy, and a chemical examination revealed massive proteinuria. His laboratory reports were as follows:

- Serum urea nitrogen: 42 mg/dl
- Serum creatinine: 3.0 mg/dl
- Serum total proteins: 4.2 g/dl
- Serum albumin: 2.2 g/ dl
- Serum globin: 2.0 g/dl

1. What is the most probable diagnosis?
2. How should the management of the patient be done?

Case Study 3

A middle-aged obese female presented with a history of epigastric pain radiating to back for 2 months, which was not related to meals. She was prescribed antacids, but she returned 1 month later with more severe pain and weight loss. She gave a history of dark urine and pale stools on examination; icterus was there, but no other abnormality was found.

1. Which biochemical investigations will you do in the serum of this patient?
2. What is your diagnosis based on lab findings and history?
3. Analyze the given urine sample and interpret.

Chapter 11

Renal Function Tests

Some routine investigations done in the laboratory during the renal function test are given in Table 11.1.

Table 11.1 Routine investigations done in the laboratory during the renal function test

Name of Investigation	Method Name	Biological Reference
Blood urea	Urease-GLDH	50 mg/dl
S. creatinine	Jaffe's kinetic	0.7–1.3 mg/dl
Uric acid	Enzymatic	3–7 mg/dl
24 h urine protein	Biuret	0.028–0.141 g/day

Urea

Urea is an end product of protein metabolism. Breakdown of proteins forms amino acids, which is precursor for the formation of ammonia (NH_3). NH_3 further get converted into urea in the liver, and the kidney eliminates it in urine.

Clinical Biochemistry: A Laboratory Guide
Rooma Devi, Aman Chauhan, Simmi Kharb, and Chandra Shekhar Pundir

ISBN 978-981-4968-75-1 (Hardcover), 978-1-003-45566-0 (eBook)
www.jennystanford.com

Methods for Urea Estimation

- Berthelot method (phenol hypochlorite method)
- Diacetyl monoxime method (DAM)
- Urease method
- Nesslerization method
- Urastrat strip
- Azostix

1. Berthelot method (phenol hypochlorite method)

Principle:

NH_3 + Phenol + Hypochlorite $\xrightarrow{\text{Nitroprusside (alk. Med)}}$ Indophenol (Blue green Complex)

λ_{580nm}

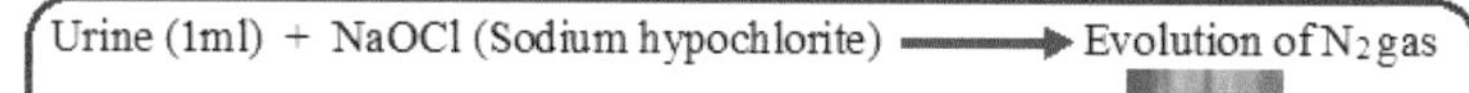

- Brisk effervescence of Nitrogen gas seen
- Indicates the presence of urea

Procedure:

Reagents	Blank	Test	Standard
Reagent 1	1 ml	1 ml	1 ml
Distilled water	10 µl	—	—
Urea standard	—	—	10 µl
Serum	—	10 µl	—
Incubate for 5 min.			
Reagent 2	1 ml	1 ml	1 ml

Incubate for 10 min, and read at 580 nm.

Calculation:

$$\text{In Urine} = \frac{\text{OD of Test}}{\text{OD of Standard}} \times \frac{\text{Amount of Standard }(\mu g)}{\text{Volume of Sample}}$$

$$\times \text{Dilution Factor} \times \frac{1000}{1000}$$

$$\text{In Serum} = \frac{\text{OD of Test}}{\text{OD of Standard}} \times \frac{\text{Amount of Standard }(\mu g)}{\text{Volume of Sample}} \times \frac{100}{1000}$$

Results = ...mg/L (Urine)
...mg/dl (Serum)

Point to be remembered: 3 ml of protein-free filtrate is equivalent to 0.3 ml serum/urine sample

Interpretation:

Normal blood urea = 10–40 mg/dl

Normal urine urea = 23–30 g/day

Normal blood nitrogen (BUN) = 7–25 mg/dl (1 mg of BUN = 2.14 mg of urea)

Advantages:

- Color stability
- Linearity

Disadvantages:

Interference from hemolysis, bilirubin, lipemic serum, and ammonia in reagents or atmosphere gives rise to false high values. So, it must be removed by adsorption on permutit or sodium aluminum silicate.

Blood Urea Estimation

Blood urea can be estimated by the diacetyl monoxime method (DAM).

Principle:

- Urea reacts with diacetyl under acidic conditions to form a colored chromogen called diazene. Yellow diacetyl is an unstable compound, so we use diacetyl monoxime.
- As the color developed is faint and unstable, ferric alum or thiosemicarbazide is used to intensify and stabilize the color.

$$\underset{\text{Urea}}{(H_2N)_2C{=}O} + \underset{\text{Diacetyl}}{H_3C{-}CO{-}CO{-}CH_3} \xrightarrow[\text{Heat}]{H^+} \underset{\text{Diazine}}{\text{Diazine}} + 2H_2O$$

Preparation of PFF:
Take 0.2 ml of oxalated blood, 1.9 ml of 5% zinc sulphate, and 1.9 ml of 0.3 N sodium hydroxide solution. Centrifuge for 5 min. The supernatant is PFF.

Procedure: Take three test tubes and label them as test (T), standard (S), and blank (B).

Table 11.2 Procedure for estimation of serum urea

	Test	Standard	Blank
PFF	2 ml	—	—
Working standard	—	2 ml	—
Distilled water	—	—	2 ml
DAM	2 ml	2 ml	2 ml
Ferric alum	2 ml	2 ml	2 ml

Keep all the test tubes in a boiling water bath for 5 min and then take OD at 420 nm.

Calculations:

$$\text{Blood urea (mg\%)} = \frac{A_t}{A_s} \times \text{Conc. of standard}$$

1 ml of standard solution has urea = 0.025 mg

2 ml of standard solution has urea = 0.05 mg

As volume of blood in 4 ml of PFF = 0.2 ml

So, volume of blood in 2 ml of PFF = 0.1 ml

$$\text{Therefore conc. of urea in 0.1 ml of blood is} = \frac{A_t}{A_s} \times 0.05$$

And conc. of urea in 100 ml of blood will be = $\frac{A_t \times 0.05 \times 100}{A_s \times 0.1}$

$$\text{i.e. Blood urea (mg\%)} = \frac{A_t}{A_s} \times 50$$

The normal blood urea level is 15–45 mg%.

Advantages:

- It is a very simple and accurate method.

Disadvantages:

- Linearity is only up to 100 mg%; after that the sample has to be diluted.
- The time for which the tubes should be kept in a boiling water bath depends on the quantity of urea, as the formation of a colored complex occurs at the boiling temperature, but we boil all the tubes for 5 min.

Importance of blood urea estimation:
Blood urea estimation is important for the assessment of kidney function since the route of urea excretion is through the kidneys.

Indirect methods (enzymatic): Urea is hydrolyzed by urease to ammonia, which is measured.

$$\boxed{NH_2CONH_2 + 2H_2O \xrightarrow{\text{Urease}} 2H_4^+ + CO_3^{2-}}$$

Nessler's Method

Ammonia generated with the assistance of the enzyme urease is made to react with Nessler's reagent (potassium mercuric iodide), which gives rise to a brown compound, which is studied at 450 nm. The enzyme acts optimally at 55°C and pH 7–8 and is inhibited by ammonia and fluoride.

Disadvantages:

- Turbidity
- Color instability
- Nonlinear calibration
- Susceptibility to contamination with ammonia from the lab and endogenous ammonia in the specimen

Urease/glutamate dehydrogenase method

In this method, first urea will get hydrolysed into ammonium ions in the presence of urease enzyme. These ammoinium ions now react with 2-oxoglutarate and form glutamate and water. This reaction is catalysed by glutamate dehydrogenase. During this reaction (catalysed by GDH), oxidation of NADH to NAD+ occurs, which is measured at 340 nm.

Other methods:

Stick tests (semiquantitative)

Urastrat strip:

Azostix:

Elevated levels of urea in blood: Increased level of blood urea is termed azotemia. When it is associated with clinical symptoms, it is called uremia. The causes of elevated urea can be divided into the following stages:

Pre-renal:

- Increased breakdown of tissue protein, e.g., in fever and other toxic states
- Dehydration leading to low perfusion of kidneys and, therefore, decreased filtration of urea in cases of:
  - Burns
  - Diarrhea
  - Vomiting
  - Severe hemorrhage
  - Diabetic ketoacidosis

Renal:

- Glomerulonephritis: acute and chronic
- Chronic renal failure
- Chronic pyelonephritis
- Poisoning, which affects the kidneys
- Nephrotoxic drugs, e.g., gentamycin, cephaloridine, sulfonamides, etc.

- Diabetic and gouty nephropathy

Post-renal:

- Stone anywhere in the urinary tract
- Malignancy
- Stricture of the urinary tract
- Prostatic hypertrophy in the elderly

The backflow to kidneys increases, and ultimately nephrons get destroyed. Increased backflow also leads to reduction in effective filtration pressure. The clinical utility of blood urea lies in its measurement in conjunction with creatinine and in distinguishing pre-renal from post-renal causes.

The normal urea nitrogen/creatinine ratio is 12–20.

(Blood urea = Blood urea nitrogen (BUN) × 2.14)

A decrease in urea levels is seen in:

- Sever liver disease
- Cancer of the liver

Physiological variations are seen during pregnancy due to hemodilution. It is low in females and varies with the amount of protein in diet.

Urea Clearance

Urea clearance is the volume of plasma cleared of urea by both kidneys in 1 min. It is calculated by the formula:

$$C_{\mathrm{m}} = \frac{uv}{p}$$

C_{m} = maximal urea clearance
u = urine urea (mg%)
v = flow of urine (ml/min)
p = blood urea (mg%)

- The formula is applicable if the output of urine is equal to or more than 2 ml/min. This is referred to as maximal urea clearance.
- The normal value is 60–95 ml/min (average = 75 ml/min).

- Urea clearance drastically changes when the volume of urine is less than 2 ml/min. So the standard urea clearance (Cs) is calculated by the formula:

$$Cs = \frac{u\sqrt{v}}{p}$$

where $\sqrt{v}$ is taken to bring the urine volume close to 1 ml/min.

The normal standard urea clearance is 40–65 ml/min (average = 54 ml/min).

Interpretation: When clearance is less than 70% of the normal, it indicates renal damage.

Grading of renal function based on urea clearance:

Over 70	Normal
70–50	Mild deficit
49–20	Moderate deficit
Below 20	Severe deficits
Below 5	Uremic coma

Creatinine

Creatinine is produced in the muscles from creatine phosphate by spontaneous dehydration. It is the waste product obtained from the catabolism of tissue proteins and is excreted by the kidneys. It is one of the most constant non-protein nitrogen present in the blood and helps in the assessment of renal function since the value of serum creatinine is entirely dependent on muscle mass and unaffected by dietary protein intake. Creatinine clearance is one of the most sensitive tests for measuring the glomerular filtration rate (GFR) (125–130 ml/min).

Method of Creatinine Estimation (Jaffe's Method)

Principle:

Creatinine + saturated picric acid $\longrightarrow$

Creatinine picrate orange coloured

- Absorbance is measured at 520 nm

Specimen:

- Blood is not used since RBCs are rich in non-creatinine chromogens, i.e., glucose, vitamin C, etc.
- Serum is preferred, but it is deproteinized before estimation.

Procedure:

Preparation of PFF:

- Take 1 ml of DW, 1 ml of serum, 1 ml of 5% sodium tungstate, and 1 ml of 2/3 N H_2SO_4. Mix at each addition and centrifuge. Take the supernatant.

Take three test tubes and label them as T (test), S (standard), and B (blank).

Table 11.3 Procedure for estimation of serum creatinine

	Test	Standard	Blank
PFF	2 ml	—	—
DW	2 ml	3 ml	4 ml
Working standard (0.01 mg/ml)	—	1 ml	—
NaOH (1.2N)	1 ml	1 ml	1 ml
Saturated picric acid	1 ml	1 ml	1 ml

Mix well and read after 15–20 min at 520 nm.

Calculations:

Serum Creatinine (mg%) = $\frac{A_t}{A_s} \times$ conc. of standard

- As conc. of standard is 0.01 mg/ml,
- Creatinine present in 1 ml of standard is 0.01 mg.
- 4 ml of PFF has serum = 1 ml
- 2 ml of PFF has serum = 0.5 ml
- 0.01 mg creatinine is present in 0.5 ml of serum,
- 100 ml of serum will have creatinine = $\frac{0.01}{0.5} \times 100$

$$\text{S. Creatinine (mg\%)} = \frac{A_t}{A_s} \times 2$$

Disadvantages:

- Lack of specificity: This reaction is sensitive to certain variables such as pH, temperature, etc.

Normal values:

Males: 0.7–1.5 mg/dl
Females: 0.4–1.2 mg/dl
It is higher in males since it is related to muscle mass.

Creatinine coefficient:

The creatinine coefficient is the mg of creatinine/kg BW/24 h urine excreted.

Males: 2–26 mg/kg BW/24 h

Females: 14–20 mg/kg BW/24 h

The creatinine coefficient is more precise and is used to assess the functioning of muscle mass and the accuracy of collection of a 24 h urine sample.

Clinical importance:

An increase in serum creatinine is seen in:

- Renal failure.
- Catabolic states, e.g., fever, dehydration, etc.
- Initial stages of muscular dystrophies.
- Crush injuries.

A decrease in serum creatinine is not significant and is seen in:

- Terminal stages of muscle-wasting disorders.

Other Methods of Estimation

1. Enzymatic Methods

(A)

$$\text{Creatinine} + H_2O \xrightarrow{\text{Creatinine hydrolase}} \text{Creatine}$$

$$\text{Creatine} + \text{ATP} \xrightarrow{\text{Creatine kinase}} \text{Creatine phosphate} + \text{ADP}$$

$$\text{ADP} + \text{Phosphoenolpyruvate} \xrightarrow{\text{Pyruvate kinase}} \text{ATP} + \text{Pyruvate}$$

$$\text{Pyruvate} + \text{NADH} + H^+ \xrightarrow{\text{LDH}} \text{Lactate} + \text{NAD}$$

(B)

$$\text{Creatinine} \xrightarrow{\text{Creatinine Iminohydrolase}} \text{N-methyl hydrantoin} + NH_3$$

NH_3 can be estimated by ion selective electrode

(C) Kodak Ektachem Analyzer

$$\text{Creatinine} \xrightarrow{\text{Deaminase}} NH_3 \longrightarrow \text{Change in colour of pH indicator}$$

2. **High Performance Liquid Chromatography (HPLC)**

Estimation of Creatinine in Urine

Urine creatinine is also estimated by using Jaffe's method. The normal urinary excretion value is 1–2 g/day

Clearance:

- It is defined as the volume of plasma cleared of a particular substance by both kidneys in 1 min.
- It acts as an indicator of GFR because after filtration, there is neither reabsorption nor secretion when the serum creatinine value is not very high.
- Any substance to be used for calculating clearance should preferably meet the following criteria:
  - It is freely filtered by the glomerulus.
  - It is neither secreted nor reabsorbed intact by the renal tubules.
  - It has a stable production rate.

Procedure:

- Adequately hydrate the patients by asking them to drink 500 ml of water.
- Any caffeine-containing compounds or drugs should be withheld, if possible, on the day of the test.
- At a defined point of time (e.g., 8 am), patients are asked to void urine and this sample is discarded.
- Rest of the urine subsequently voided over a prescribed period of time is then collected.
- A blood sample is collected in between.

Calculation:

- Creatinine is estimated in both the samples, and then its clearance is calculated using the following formula:

$$\text{Clearance (ml/min)} = \frac{uv}{p}$$

where u = urine creatinine (mg%)
v = flow of urine (ml/min)
p = serum creatinine (mg%)

- The normal creatinine clearance in males is 85–125 ml/min.
- The normal creatinine clearance in females is 80–115 ml/min.

Significance:

- Decreased creatinine clearance is a very sensitive indicator of decreased GFR and can help to detect renal failure in the initial stages when serum creatinine is still in the normal range.
- Clearance values up to 75% of the average normal value may indicate adequate renal function.

Multiple Choice Questions

1. Conversion of creatine to creatinine is
   a. An enzymatic process
   b. Non-enzymatic process
   c. Both of the above
   d. None of the above
2. The most specific method for estimating serum creatinine is
   a. Enzymatic method
   b. Jaffe's reaction
   c. Both of the above
   d. None of the above
3. The conversion of creatine to creatinine involves
   a. Dehydration
   b. Cyclization
   c. Both of the above
   d. None of the above

Case Studies

Case Study 1

A 12-year-old kid is delivered to his specialist with complaints of painless swelling of each leg, which was noted around 2 months ago. The swelling started at the ankles; however, currently his legs, thighs, and sex organ are swollen. His face is puffy in the mornings on waking up. His weight increased by 10 kg over the past 3 months. His urine seems frothy. He has noticed increasing shortness of breath but denies any pain. He has conjointly developed spontaneous bruising over the past 6 months.

1. Which biochemical investigations will you do in the serum of this patient?
2. What is your diagnosis based on the lab findings and history?
3. Analyze the given urine sample and interpret.

Case Study 2

A 45-year-old man complained of finger joint inflammation and severe pain. His laboratory test reports are as follows:

Serum uric acid: 8.7 mg/dl
RA test: Negative

1. What is your probable diagnosis?

Case Study 3

A 34-year-old diabetic female was advised by her physician to provide her blood specimen for "lipid profile tests." The reports are as follows:

Blood glucose fasting: 235 mg/dl
Serum total cholesterol: 286 mg/dl
Serum triglycerides: 266 mg/dl
Serum LDL cholesterol: 166 mg/dl
Serum VLDL: 49 mg/dl
Total serum cholesterol/HDL-cholesterol: 5.4
LDL-cholesterol/HDL-cholesterol: 3.9

1. What is your probable diagnosis?

Chapter 12

Pancreatic Function Test

The pancreas is 12–15 cm long and lies across the posterior wall of the abdominal cavity. Its head is found within the curve of the small intestine, and the body and tail are directed toward the left, extending to the spleen. Pancreatic digestive enzymes containing bicarbonate-rich juice enter the small intestine. It is located behind the peritoneal cavity.

Function of Pancreas

Exocrine: Due to digestion of food

Endocrine: Due to hormone secretion

Pancreatic Fluid

- 1.5 to 2 L/day fluid is secreted
- Rich in digestive enzyme
- Fluid secretion by acinar cells
- Pancreatic fluid is clear, colorless, and watery (pH 8.3)
- Due to H_2CO_3
- Neutralizes HCl from gastric juice
- **Enzymes**
  - Proteolytic trypsin, chymotrypsin, elastin collagenase, amino peptides, carboxypeptidase

Clinical Biochemistry: A Laboratory Guide
Rooma Devi, Aman Chauhan, Simmi Kharb, and Chandra Shekhar Pundir

ISBN 978-981-4968-75-1 (Hardcover), 978-1-003-45566-0 (eBook)
www.jennystanford.com

- Lipase
- Amylase
- Nuclease

Utility of Assessment of Functions of Pancreas

- Pancreatitis
- Chronic diarrhea
- Malabsorption, maldigestion
- Cystic fibrosis
- Pancreatic carcinoma
- Primary or secondary pancreatic insufficiency syndromes
- Anatomical and morphological disorders of pancreas

Test of Pancreatic Function Test

1. **Tests for detection of malabsorption**
   - Stool examination for extra fat
   - D-xylose test
   - Fecal fat analysis
2. **Tests for exocrine function**
   - Secretin
   - CCK
   - Fecal fat
   - Trypsin
   - Chymotrypsin
3. **Test to assess extrahepatic obstruction**
   - Bilirubin
4. **Endocrine-related tests**
   - Gastrin
   - Insulin
   - Glucose
5. **Imaging/noninvasive tests**
   - USG
   - Radiograph
   - MRI
   - ERCP

Tests for Exocrine Function

Invasive tests: Gastrointestinal fluid aspiration by incubation is required. After an overnight fasting, a stimulus is given to the pancreas, followed by the measurement of:

- The total volume of pancreatic juice
- The concentration of secreted carbonate
- The activities of duct gland enzymes

The following tests were designed on the bases of principle:

Exocrine Function

1. **Lundh's test**
   - After giving a standard meal (obsolete now)
   - A standard meal consists of 5% protein, 6% fat, 15% carbohydrate, and 74% non-nutrient fiber
   - This test prevents the determination of total enzyme, HCO_3, or secretory volume
2. **Secretin stimulation test**
   - Overnight fasting
   - Basal samples collected from stomach and duodenum
   - 0.25 to 1 U/kg secretin is administrated (1/v)
   - Duodenal fluids collected at 15 min interval for 1 h
   - Secretin-induced stimulation of pancreatic enzyme is inconsistent
3. **Secretin–cholecystokinin stimulation test**
   - Overnight fasting
   - Basal fluid collected
   - 1 U/kg of secretin is administrated (IV) followed by administration of 30 ng/kg of caerulein (IV).
   - Caerulin synthetic decapeptide of CCK, which stimulates pancreatic enzyme secretion
   - Pancreatic juice collected every 10–15 min
   - Volume pH, HCO_3^-, enzyme are measured
4. **Fecal fat**
   - Fecal fat derived from unabsorbed ingested lipid
   - Lipids excreted in intestine
   - Metabolism of intestine bacteria

Qualitative screening: Using Sudan III dye

Quantitative fecal fat analysis

72 h sample to 5-day sample

Methods:

- Gravimetric
- Titrimetric
- IR magnetic resonance
- NMR

5. **Amylase and lipase**

   Amylase is the main enzyme for pancreatic function.

   - Acute pancreatitis
   - Reduces peak in 24 h and because of its clearance by the kidneys, it returns to normal in 3–5 days, often making urinary amylase a more sensitive indicator

$$\% \frac{\text{Amylase Clearance}}{\text{Creatinine}} = 100 \times \frac{\text{Urine Amylase}}{\text{Serum Amylase}} \times \frac{\text{Serum Creatinine}}{\text{Urine Creatinine}}$$

Reference value:

- Normal: <3.1%
- Pancreatitis: >8% or 9%
- Increases in burns, diabetic ketoacidosis (DKA), and sepsis

Lipase

- More sensitive than amylase
- Increased level for 7–14 days
- Increases in bone fracture associated with fat emboli

Fecal elastase-1

- Chymotrypsin-like enzyme
- Secreted by pancreas
- Sensitive test
- Useful in the diagnosis of CF

Multiple Choice Questions

1. Salivary amylase breaks down starch into
   (a) Fructose

(b) Maltose
(c) Glucose
(d) Galactose

Ans: b

2. Pancreas is a specialized organ that acts as
   (a) Endocrine gland
   (b) Exocrine gland
   (c) None of above
   (d) Both (a) and (b)

Ans: d

Chapter 13

Acids and Bases

Acid: An acid can be defined as a proton donor that increases the concentration of hydrogen ions in a solution.

$$HCl \leftrightarrow H^+ + Cl^-$$

Base: A base can be defined as a proton acceptor, electron pair, or hydroxide ion donor that decreases the concentration of hydrogen ions in a solution.

$$KOH \leftrightarrow K^+ + OH^-$$

pH and Its Measurements

What is pH?

pH is the measurement of the concentration of hydrogen ions (H^+) present in a dilute solution. It is the measure of the level of acidity or basicity of a solution.

Formula and Its Scale

The formal definition of pH is the negative logarithm of the hydrogen ion activity:

$$pH = -\log[H^+]$$

It is measured on a scale of 0 to 14. For example, in a solution, if the H^+ ion concentration is 10^{-6} M, then the pH is 6.

Clinical Biochemistry: A Laboratory Guide
Rooma Devi, Aman Chauhan, Simmi Kharb, and Chandra Shekhar Pundir

ISBN 978-981-4968-75-1 (Hardcover), 978-1-003-45566-0 (eBook)
www.jennystanford.com

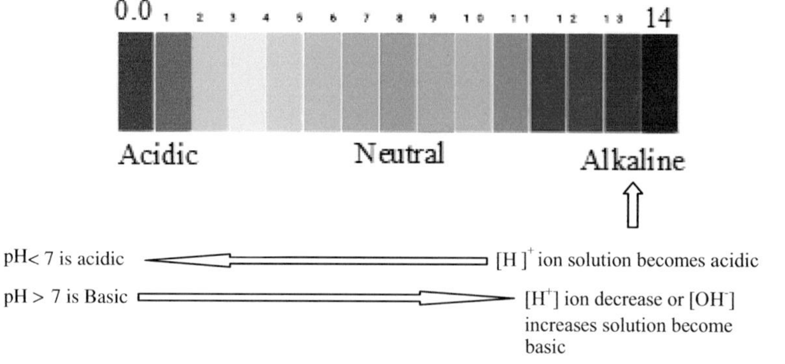

Measuring the pH of a Solution

pH Paper

- Shows pH at intervals of 1 (or 0.5)
- Cannot give more precise values such as pH 8.2, 6.8 (low accuracy)
- Not temperature sensitive

Advantages:

- Cheaper and easily available
- Gives rough estimation

pH Meter

- Can give precise pH values up to two decimal places (3.55, 7.42)
- Temperature sensitive
- More accurate values than pH paper
- More expensive

The pH measurement system consists of

- A pH-measuring electrode
- A reference electrode
- A high input meter

The pH-measuring electrode is a glass electrode, which utilizes a thin glass membrane responsive to change in H^+ activity.

How does it work?

- A voltmeter measures the difference between the voltages of the two electrodes.
- The meter then translates the voltage difference into pH and displays it on the screen.

pH Meter Calibration

Some factors that may lead to systematic errors in pH measurement results are as follows:

- A symmetric potential of glass electrode
- Aging and coating of the glass electrode over time, leading to inaccuracies
- Clogging of glass electrode diaphragm

Standardization: A standard is used to calibrate analytic instruments.

Before taking a pH measurement, it is essential to calibrate the pH using standard solutions of known pH. This standardization is necessary to maintain a reference state of pH.

Why is pH important?

- pH affects the solubility of many substances.
- pH affects the structure and function of most proteins, including enzymes.

Buffer

A buffer is a combination of either a weak acid and its salt or a weak base and its salt. A buffer solution can resist changes in pH when small amounts of acids or bases are added to it.

Why are buffers important?

- Buffers regulate the pH of body fluids and tissues.
- The physiological pH of blood is 7.4.

The choice of a buffer solution is important before starting any biochemistry experiment. A buffer maintains the hydrogen ion concentration of a solution by absorbing protons throughout the reaction or by releasing protons once they are consumed throughout the reaction.

Buffer Capacity

The buffer capacity is measured by the power of the buffer to retain its hydrogen ion concentration once strong acids or bases are added to the medium. In other words, the buffer capacity corresponds to the number of H^+ or OH^- ions that may be neutralized by the buffer.

Preparation of Buffer Solutions

1. Titration Method

In the titration method, one of the two compounds (acid/base) is added slowly to the other solution until the required pH is obtained.

Advantages

- Easy to understand
- Useful when one form of buffer is available

Disadvantages

- Slow
- May require a lot of base (or acid)

2. Calculation from pKa (H–H Equation)

The Henderson–Hasselbalch equation is as follows:

$$pH = pKa + \log (base/acid)$$

For a particular pH, choose a buffer having a pKa value close to the desired pH. The calculation will give the ratio of the base and acid that should be mixed to obtain the final pH value.

Advantages:

- Process is fast, and the buffer can be prepared easily.
- pH adjustment is rarely necessary.

Disadvantages:

- The pKa of buffer must be known before preparation of buffer.

3. Refer to a Published Table

The essential characteristics for a suitable biological buffer are as follows:

- pKa between 6 and 8, water soluble, chemically stable, biochemically inert
- Does not form a complex with metal ions
- Minimal effect of salt
- Should not absorb light in the UV and visible range
- Can be easily prepared

Commonly Used Buffers

Phosphate Buffers

- They have a useful pH range of 6.5 to 7.5 and are widely used.
- They are preferred because they are naturally present in cells and hence provide a more physiologically similar environment than other synthetic buffers.
- Acid: Monosodium hydrogen phosphate
- Base: Disodium hydrogen phosphate

Advantages

- They have a very high buffering capacity.
- They are highly soluble in water.

Disadvantage

- They might inhibit enzymatic reactions and procedures.
- They precipitate in ethanol. So, they cannot be used to precipitate DNA or RNA.

Multiple Choice Questions

1. Which of the following describes a universal property of buffers?
   a. Buffers are usually composed of a mixture of strong acids and strong bases.
   b. Buffers work best at the pH at which they are completely dissociated.

c. Buffers work best at the pH at which they are 50% dissociated.

d. Buffers work best at one pH unit lower than the pKa.

e. Buffers work equally well at all concentrations

Ans: e

2. Buffer solutions resist any change in pH. This is because

   a. Of the fixed value of pH

   b. They give unionized acid or base on reaction with added acid or alkali

   c. Of an excess of H^+ or OH^- ions

   d. Acids and alkalis in these solutions are shielded from attack by other ions

Ans: b

Practical Spots in Biochemistry

Spot 1

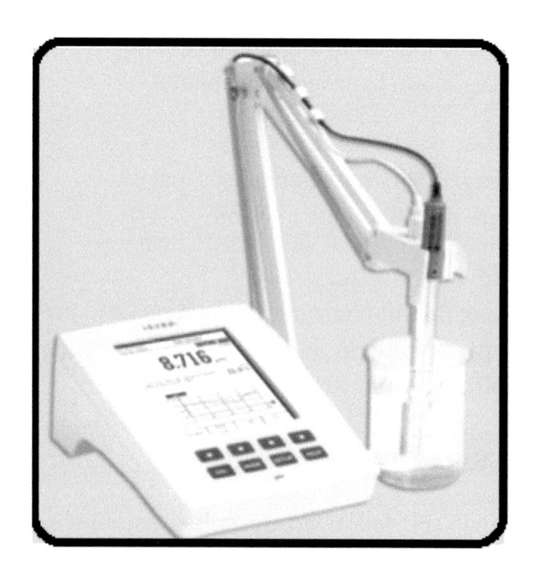

1. Identify the instrument.
2. Add a note on its principle and clinical applications.

Chapter 14

Colorimetry

Photometry

Photometry is the measurement of the propagation of light through various materials. The color of light is a function of its wavelength and a change in the color of light can be detected in the visible region with a change in wavelength.

In photometry, measurement of the color intensity of a solution is governed by two laws:

1. Beer's law
2. Lambert's law

Table 14.1 Colors of the visible spectrum absorbed at different wavelengths and their corresponding colors of solution

Wavelength (nM)	Color absorbed (filter)	Color of solution
400–435	Violet	Green-yellow
435–500	Blue	Yellow
500–570	Green	Red
570–600	Yellow	Blue
600–630	Orange	Green-blue
630–700	Red	Green

Clinical Biochemistry: A Laboratory Guide
Rooma Devi, Aman Chauhan, Simmi Kharb, and Chandra Shekhar Pundir

ISBN 978-981-4968-75-1 (Hardcover), 978-1-003-45566-0 (eBook)
www.jennystanford.com

Colorimetry

Colorimetry is the measurement of color. It measures the concentration of substances that are colored or can be converted into colored compounds by a suitable reaction. This technique is very sensitive and requires minute amount of samples.

A colorimeter is an instrument used to measure the concentration of colored substances. It works in the visible range of 400–800 nm of the electromagnetic spectrum of light.

Principle: When white light passes through a colored solution, some light is absorbed and some light is transmitted. The absorbed light is measured as optical density (OD). This absorbed light is made to fall on a photo cell, which converts light energy into electrical energy, which is measured by a galvanometer.

In colorimetry, it is preferable to use monochromatic light of specific wavelength, which is absorbed maximally by the chromogen being measured. The relationship between the absorption (or transmission) of light and the concentration of chromogen was described by Beer, and that between the absorption of light and the thickness of the solution was described by Lambert.

Beer's Law

The amount of light absorbed by a colored solution is proportional to the concentration of the solution. If A is the light absorbed (absorbance) and C is the concentration of the coloring substance, then $A \infty C$.

The law states that the log of the ratio of the intensities of incident light (I°) and emergent light (I) is directly proportional to the concentration of the coloring substance (C) in the solution, provided the thickness of the solution through which the light passes in constant (K). That is,

$$\text{Log } I^{\circ}/I = KC$$

Lambert's Law

This law states that the log of the ratio of the intensities of incident light (I°) and emergent light (I) is directly proportional to the thickness of the solution (t) through which the light passes provided that the concentration of the chromogen is constant. That is,

$$\text{Log } I°/I = K2t$$

The amount of light absorbed by a colored solution is proportional to the depth through which the light passes in the solution. If L is the depth through which light passes in the solution, then $A \propto L$.

Beer–Lambert Law

The Beer–Lambert law states that the log of the ratio of the intensities of incident light (I°) and emergent light (I) is directly proportional to the concentration of the chromogen and the thickness of the solution through which light passes. That is,

$$\text{Log } I°/I = KCt$$

The ratio (I/I°) of the intensities of emergent light and incident light is known as transmittance (T). It is the measure of the ability of a solution to transmit light.

$$\text{Log } 1/T = KCt$$

Log 1/T is known as the optical density (OD) or absorbance (A). It is a measure of the ability of a solution to absorb light.

$$A = KCt$$

and

$$A \propto Ct$$

Transmittance (T) has a logarithmic relationship and absorbance (A) has a direct relationship with C and t. Therefore, absorbance is used in calculations in all the colorimeter determinations.

Table 14.2 UV, visible, and infrared spectrum characteristics

Wavelength (nm)	Region Name	Color Observed/ Complementary Color
<380	UV	Not visible
380–440	Visible	Violet (green-yellow)
440–500	Visible	Blue (yellow)
500–580	Visible	Green (blue)
580–600	Visible	Yellow (blue)
600–620	Visible	Orange (green-blue)
620–750	Visible	Red (green)
>750	Infrared	Not visible

The basic components of a colorimeter are as follows:

1. **Source of light:** Source of light in visible range may be an electric lamp such as tungsten lamp and in UV range, the source of light may be deuterium or hydrogen lamp.
2. **Filters:** The light from the electric lamp is passed through a filter. Filters are usually made up of colored glass or dyed gelatin. A set of filters is provided with the instrument. For each determination, a suitable filter should be selected.
3. **Photocell or phototube:** It is used to convert the transmittance light into electrical light (readable form). By passing the dispersed light through a narrow silt, monochromatic light of specific wavelength can be obtained.
4. **Measuring device:** The potential difference is measured by a galvanometer. The galvanometer is generally calibrated to directly read transmittance or absorbance or both.

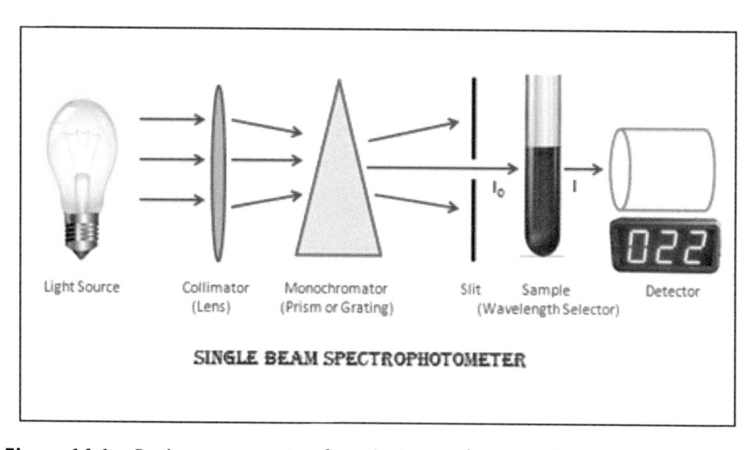

Figure 14.1 Basic components of a colorimeter/spectrophotometer.

Spectrophotometer

The spectrophotometer is based on the principle of Beer–Lambert law and is used to measure the amount of light absorbed by the sample.

- With the help of prism or optical device, a light of specific bandwith is selected.

- For measurements within the ultraviolet and infrared, it is not essential that the analyte should show a color even if it absorbs light in these wavelengths.
- A photometer does not have a filter; instead, it has a prism or optical device with a slit.
- The photometer uses monochromatic light and is more expensive than a filter photometer due to its increased sensitivity, spectral variation, and narrowness of the spectral region isolated.

Spectrophotometers are of two types: single beam and double beam.

- In single beam spectrophotometers, a single beam of light is used. The reference sample is placed in sample holder and the absorbance value is measured. The sample to be measured is then placed in the sample holder and again the absorption rate is measured.
- The absorbance value of the reference sample is deducted from the absorbance value of samples to be analysed to remove the additive effects from the solvent as well as the cell.
- In a double beam photometer, the light beam from the source is split into two beams.
- One beam will pass from the reference, or the blank, and the other beam passes through the sample into consideration. Both beams recombine before reaching the monochromator. In some cases, two monochromators are also used.

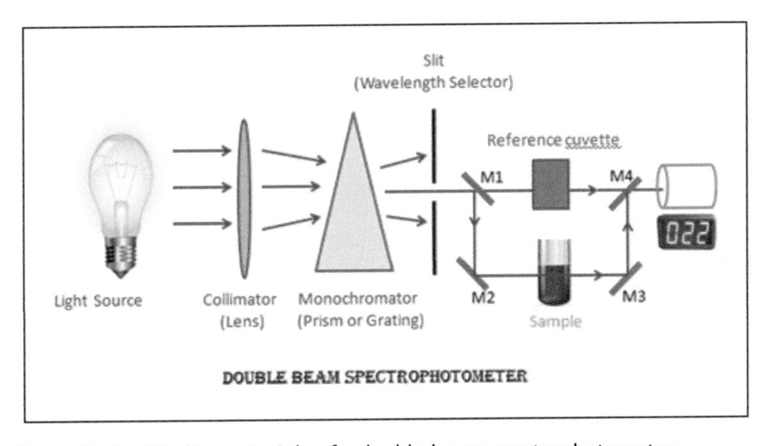

Figure 14.2 Working principle of a double-beam spectrophotometer.

- In these instruments, output readings of two to six samples can be taken in a single run.

Multiple Choice Questions

1. The Beer–Lambert law gives the relation between which of the following:
   a. Concentration and reflected radiation
   b. Concentration and scattered radiation
   c. Reflected radiation and energy absorption
   d. Energy absorption and concentration

Ans: d

2. Beer's law states that the intensity of light decreases with respect to
   a. Distance
   b. Concentration
   c. Composition
   d. Volume

Ans: a

Chapter 15

Chromatography

Chromatography is a biophysical technique, which allows the separation and purification of a molecule of interest from a crude mixture for quantitative and qualitive analysis. The term "chromatography" was given in 1906 by the Russian botanist Mikhail Tsvet.

Principle

- Components of chromatography: Mobile phase, stationary phase, and a crude sample.

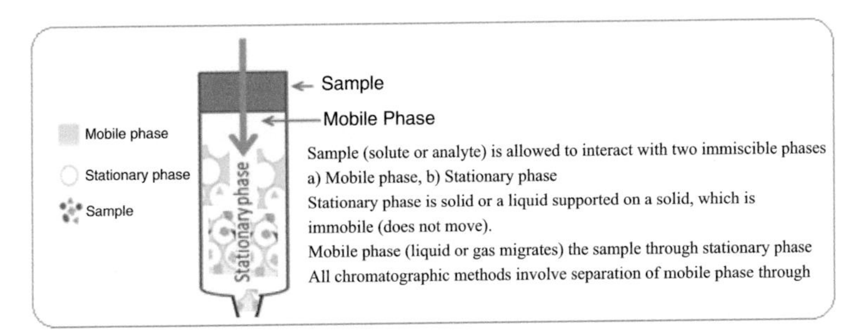

Clinical Biochemistry: A Laboratory Guide
Rooma Devi, Aman Chauhan, Simmi Kharb, and Chandra Shekhar Pundir

ISBN 978-981-4968-75-1 (Hardcover), 978-1-003-45566-0 (eBook)
www.jennystanford.com

Table 15.1 Chemical and physical properties on the basis of which molecules are separated

Physical Properties	Chemical Properties
1. Molecular weight (MW) 2. Freezing point 3. Boiling point 4. Crystallization 5. Solubility 6. Density	1. Functional group 2. Reactivity toward another reagent to form complex

Mechanism of Separation

- Partitioning
- Adsorption
- Exclusion
- Ion exchange
- Affinity

Classification of Chromatography

Based on Partitioning Effect (Mechanism)

- Partition chromatography
- Adsorption chromatography

Table 15.2 Classification of chromatography based on partitioning effect

Partition Chromatography	Adsorption Chromatography
An analyte distributes itself into two phases: stationary and mobile. The separation is based on solute partitioning between two liquid phases (relative solubility).	Matrix molecules have the ability to hold the analyte on their surface through a mutual interaction due to different types of forces such as hydrogen bonding, electrostatic interaction, van der Waals, etc.
Advantages: Simple, low cost, and broad specificity	**Advantage:** Retains and separates some compounds that cannot be separated by others, e.g., separation of geometrical isomers.

Partition Chromatography	Adsorption Chromatography
Disadvantages: It is used to separate only soluble analytes of sample mixture and sometime high volume of mobile phase is required for separation.	**Disadvantages:** Very strong retention of some solutes, may cause catalytic changes in solutes
Examples: Paper chromatography, thin layer chromatography, cellulose, starch, or silica matrix	**Examples:** Affinity chromatography, ion exchange chromatography, hydrophobic interaction chromatography

Based on Phases

1. **Solid phase chromatography**
   - Solid–liquid chromatography
   - Solid–gas chromatography
2. **Liquid phase chromatography**
   - Liquid–liquid chromatography
   - Liquid–gas chromatography
3. **Based on the shape of stationary phase**
   - **Planner chromatography**
     Paper chromatography
     Thin layer chromatography
   - **Column chromatography**
     - Packed column chromatography, e.g., gas chromatography, HPLC
     - Open tubular chromatography

Based on the chemical nature of the stationary phase and mobile phase:

- **Normal phase chromatography:** The stationary phase is polar in nature, while the mobile phase is nonpolar.

Reverse-phase chromatography

The stationary phase in reverse phase HPLC is nonpolar, while the mobile phase is polar.

Depending on the purpose of chromatography experiment:

- **Preparative chromatography:** It is also applied exclusively in column chromatography.
- **Analytic chromatography:** A very small amount of sample is applied or injected, and the purpose is identifying the components in the sample and also their individual concentrations.

Depending on the chemical or physical properties of stationary phase:

- High-performance liquid chromatography
- Gas chromatography
- Column chromatography
- Ion exchange chromatography
- Size exclusion chromatography
- Affinity chromatography
- Supercritical fluid chromatography
- High-performance thin-layer chromatography
- Liquid chromatography

Partition Chromatography

Paper Chromatography

- In a chromatography, separation of the mixture is performed on a paper strip which may be a stationary phase and a liquid solvent act as a mobile phase.
- High quality absorbent paper (Whatman No.1 or 3) is employed because the stationary phase.
- A solution of the sample is formed up and a really small spot is placed onto one end (usually 2cm) above a strip of the paper with a capillary.
- The position of the spot is named the origin.
- The paper is then placed in a container in order that the sting of the paper below the spot is submerged during a solvent mixture consisting butanol, acetic acid, and water in 4:1:5 ratio for separation of amino acid.
- Water held back by paper is called the stationary phase. Organic solvent is called mobile phase.

- In ascending type chromatography, the solvent flows past the spot of application by capillary action. In descending chromatography, the solvent moves downcast.
- As the solvent (mobile phase) rises up the paper, the components of every sample separate.
- Amino acids are soluble more in organic phase that moves faster. After a chromatography run in closed chamber for several hours, the paper is taken out, dried and sprayed with ninhydrin solution.
- Several purple spots develop, each representing one amino acid (Ninhydrin forms purple complex with alpha amino acid).
- Branched chain and aromatic amino acids move fastest, while acidic and basic amino acids show least mobilities.
- The chemical nature of individual spots is often identified by running known standards samples with the unknown mixture of sample.

The migration of a substance is usually expressed by the retention factor (Rf).

$$Rf = \frac{\text{Distance travelled by substance}}{\text{Distance travelled by solvent}}$$

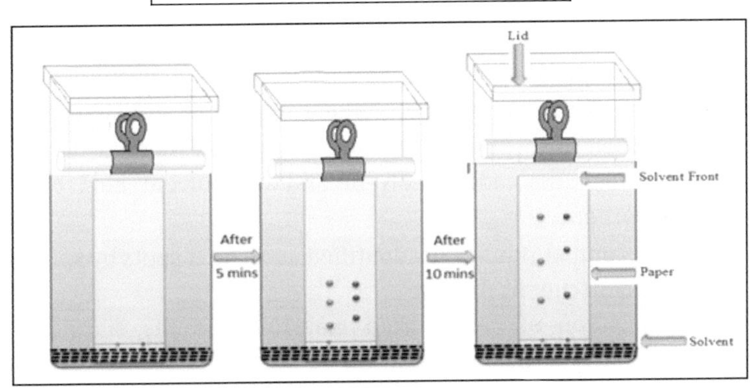

Figure 15.1 Amino acid separation by paper chromatography.

Advantages:

- Easy to handle and store
- Cheap

- Little preparation requires
- More efficient for polar and water-soluble compound

Thin-Layer Chromatography

- The principle of thin-layer chromatography (TLC) is the same as that of chromatography (partition). It also separates and analyzes complex biological and non-biological samples into their constituents.
- It is most popular for the estimation of a substance in a mixture and monitoring the progress of a chemical reaction.
- It is also one of the favored techniques for testing the purity of a sample.
- In this technique, silica gel will act as stationary phase which is to be coated as a thin layer on glass or aluminium foil. Sample is applied on lower end as small spot and then mobile phase is allowed to pass through this stationary phase.
- The sample is applied as a small spot at the lower end of the plate, which is positioned in a close glass chamber containing a solvent system (benzene–acetone mixture).
- The solvent ascends through the adsorbent due to capillary action, carrying the solutes to different distances.
- In comparison to other chromatographic techniques, the mobile phase runs from the bottom to the top by diffusion.
- The relative mobilities of the compounds are determined by their adsorption in the matrix and also by their partition between the mobile solvent and the solvent held by the adsorbent.
- The compounds can be identified as distinct spots by spraying suitable chemical agents.

The position of immobilized molecules on the silica gel is controlled by multiple factors:

- Functional group present on the biomolecules
- Functional group present on the stationary phase
- Composition of the mobile phase
- Thickness of the stationary phase

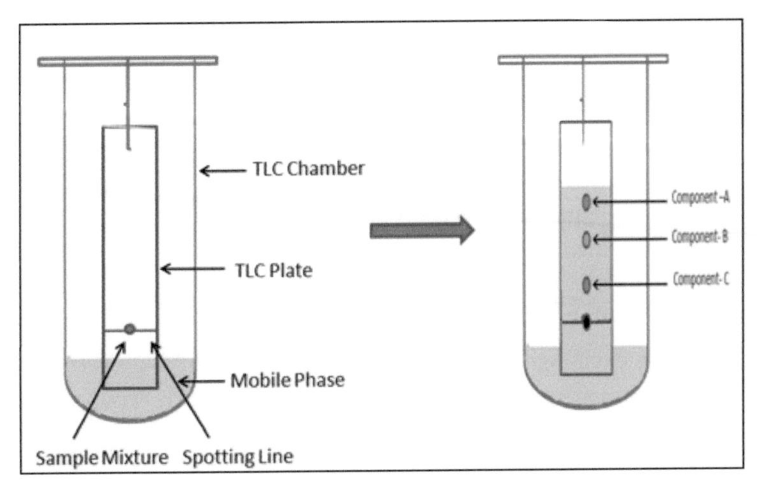

Figure 15.2 Thin-layer chromatography.

Applications of TLC

- Composition analysis/synthetic preparation of biomolecules
- Identification of impurities in the sample
- Quality testing of compounds
- Progress of chemical reactions
- Biomolecule estimation
- Bioassay

Advantages:

- Faster
- Detects smaller amounts of samples
- A wide range of stationary phases available
- Corrosive materials can be used
- Better separation of less polar compounds

Adsorption Chromatography

- In adsorption chromatography, the absorbent is used as the stationary phase.
- Silica gel, calcium hydroxyapatite, alumina, and charcoal are examples of adsorbents, which are packed into a column of glass tube.

- The sample mixture in the solvent is loaded in the column.
- The individual components get differently adsorbed on the adsorbent.
- The individual components come out of the column at different rates, which may be separately collected and identified.
- For instance, amino acids and proteins can be analyzed by the ninhydrin colorimetric method.

Ion Exchange Chromatography

- Ion exchange chromatography is a versatile, high-resolution chromatographic approach to purify proteins from a crude mixture sample.
- Separation of charged molecule is done by this type of chromatography, e.g., protein and amino acids. Column is filled with immobilized charged polymer beads either positively charged beads (termed as anion exchanger) or negatively charged beads (termed as cation exchanger).
- Negatively charged proteins bind to positively charged beads and vice versa.
- Bound protein is eluted with salt.
- The least charged protein is eluted first.
- Commercially available ion exchange resins are used.
- Ion exchange resins consist of a support mixture to which cationic and anionic groups are covalently attached.
- **Cation exchange chromatography:** A matrix with negatively charged functional groups has an affinity for positively charged molecules. The positively charged analyte displaces the reversibly bound cation and binds to the substrate.
- **Anion exchange chromatography:** A matrix carrying a fixed positively charged functional group has an affinity for negatively charged molecules. The negatively charged analyte replaces the reversibly bound anion and binds to the matrix.
- In the separation of amino acids, strong ion exchange resins based on the matrix of polystyrene cross-linked with divinylbenzene are used.
- Individual amino acids emerge at different stages, which can be collected as fractions. The order of elution of amino acids is as follows:

- Serine, aspartic acid, glutamic, threonine, and proline at pH 3.25.
- Alanine, valine, leucine, isoleucine, glycine, cysteine, methionine, tyrosine, and phenylalanine at pH 4.25.
- Lysine, histidine, and arginine at pH 5.28.

- Individual amino acids in the collected fractions can be estimated by ninhydrin/dansyl chloride reaction.

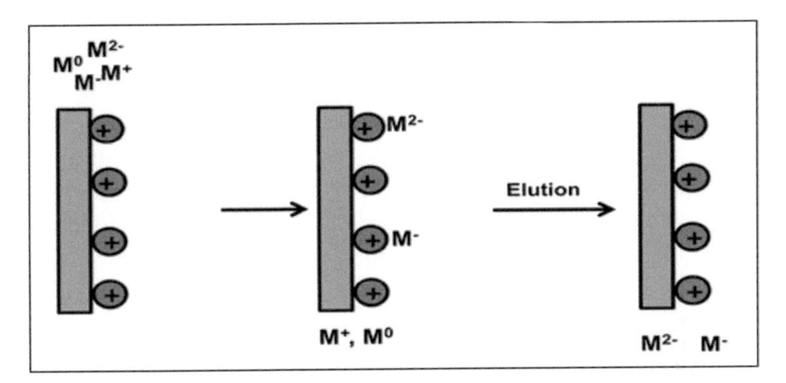

Figure 15.3 Affinity of analytes (M, M^+, M^{-1}, M^{-2}) toward a positively charged matrix.

Table 15.3 Ion-exchange matrix list

S. No.	Type of Ion Exchanger	Name	Functional Group
1.	Anion	Diethylaminoethyl (DEAE)	$-OCH_2CH_2NH\ (C_2H_5)_2$
2.	Anion	Quaternary aminomethyl (Q)	$-OCH_2N(CH_3)_3$
3.	Cation	Carboxyl methyl (CM)	$-OCH_2COOH$
4.	Cation	Sulphopropyl (SP)	$-OCH_2CH_2CH_2SO_3H$
5.	Cation	Sulphonate (S)	$-OCH_2SO_3H$

Hydrophobic Interaction Chromatography

- Hydrophobic interaction chromatography (HIC) exploits the power of a robust interaction between the hydrophobic group attached to the matrix and the hydrophobic group present on an analyte such as protein. This technique is initially termed "salting out."

- The salt within the buffer reduces the solvation of sample solutes. As solvation decreases, the hydrophobic region that becomes exposed is absorbed by the media.
- The more hydrophobic the molecule, the lower the amount of salt required for pushing the binding.
- The addition of a lower amount of salt to the protein solution leads to the displacement of the bonded water molecule with a rise in protein solubility. This effect is termed "salting in".
- In the presence of a greater amount of salt, the water molecules shielding the protein side chains are completely displaced with an exposure of hydrophobic amino acids on the protein surface to induce protein precipitation or a decrease in protein solubility. This effect is termed "salting-out."
- The exposure of hydrophobic amino acids facilitates the binding of protein to the nonpolar ligand attached to the matrix. When the concentration of salt is decreased, the exposed hydrophobic patches on the protein reduce the affinity toward the matrix and as a result it gets eluted.

Gel Filtration/Molecular Exclusion/Gel Permeation Chromatography

- This technique separates analytes on the basis of molecular size. The term "gel filtration" was given by Porath and Flodin.
- In this type of chromatography, sample (e.g., mixture of proteins) are allowed to pass through a column (stationary phase) which is filled with hydrated porous beads.
- The beads utilized in gel filtration chromatography are formed from cross-linked materials (such as dextran in Sephadex) to make a 3D mesh.
- The 3D mesh swells within the mobile phase to develop pores of various sizes. The extent of cross linking controls the pore size within the gel beads.
- The technique is used for determining the size of proteins and also for separating proteins based on molecular weight.

- When proteins are passed through a column packed with cross-linked hydrophilic polymers of arylamide, agarose, or dextran, they elute in the order of decreasing size.
- By varying the degree of cross linking, the range of sizes in which separation takes place can be fixed anywhere between 10^3 and 10^7 Daltons.
- When a large molecule is allowed to flow through the column bed, it does not penetrate the polymer matrix. Thus, it moves faster and has a short path length through the column.
- On the other hand, a small molecule permeates through the polymer gel and travels a longer path before coming out.
  V_t = Total volume occupied by the whole column
  V_o = Void volume (volume excluding the volume of polymer gel)
  V_e = Elution volume of a particular small molecule
- The elution position of molecules differing in size is given by the following equation:

$$K = \frac{V_e - V_o}{V_t - V_o}$$

- A plot of V_e/V_o against the log of molecular weights gives a straight line, which can be used for determining the molecular weight of unknown proteins once V_e is determined. This is done by employing a calibrated column with substances of known molecular weight.

Table 15.4 Gel filtration matrix list according to molecular weight

S. No.	Name of the Matrix	Molecular Weight (Daltons)
1.	Sepharose (6B)	10000–4000000
2.	Sepharose (4B)	60000–20000000
3.	Sephadex (G200)	5000–600000
4.	Sephadex (G100)	4000–150000
5.	Sephadex (G50)	1500–30000
6.	Sephadex (G25)	1000–5000
7.	Sephadex (G10)	Up to 700

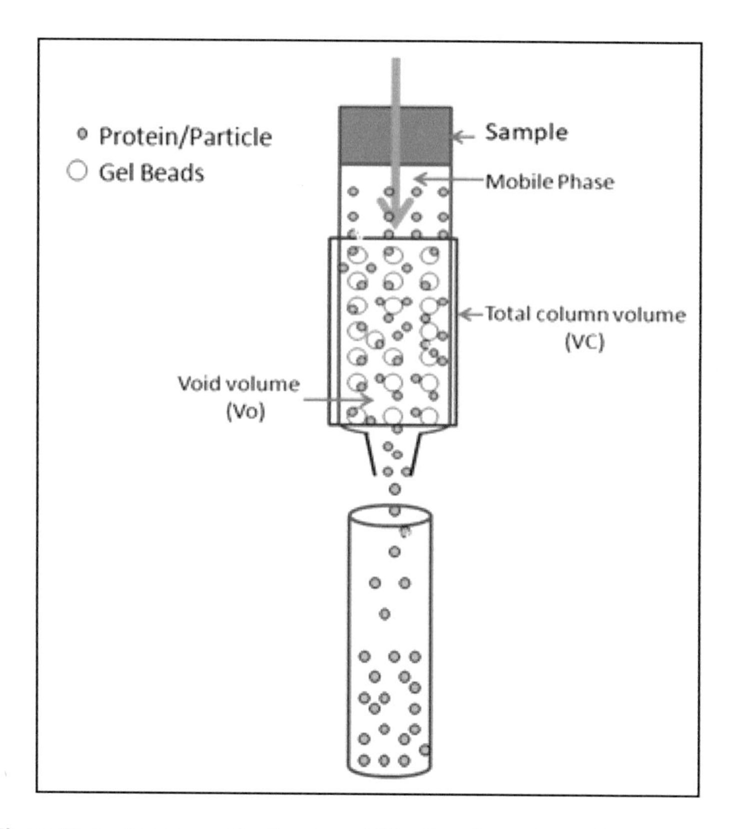

Figure 15.4 Protein purification by gel filtration chromatography.

Principle:

- The separation between molecules occurs because of the time they travel while coming out of the pores.
- When the mobile phase goes through the column, it takes protein along with it.
- The tiny molecules present in the inner part of the gel take a longer flow of liquid (or time) and travel a longer path to come back, whereas larger molecules travel a shorter distance to return.

Applications:

- Gel filtration chromatography determines the native molecular weight of a protein.

Affinity Chromatography

- Affinity chromatography is a powerful chromatographic technique used in the isolation of proteins, vitamins, enzymes, nucleic acids, drugs, antibodies, and hormone receptors.
- It is based on the principle of mutual recognition of forces between a receptor and a ligand.
- In this procedure, a specific ligand that has a high affinity for a particular protein is covalently attached to an insoluble matrix, such as crosslinked agarose.
- The mixture containing the protein to be isolated is allowed to permeate through the column containing the affinity matrix. All other proteins come out of the column after washing.
- The protein is then desorbed by eluting with a substance that weakens the interaction between the ligand and the protein; for example, the dehydrogenase enzyme specially bounds to Cibacron Blue at pH 7.0 when it is immobilized. The bound enzymes can be eluted with substances such as NADH or NADPH.
- In a clinical biochemistry laboratory, a matrix containing boronic acid is employed to separate and quantify glycosylated hemoglobin from the blood of a diabetic patient.
- An affinity matrix containing boronic acid is also used to detect ribonucleoside in the urine samples of patients.
- The avidin–biotin complex is used to isolate cytokines from immune cells (Fig. 15.5).

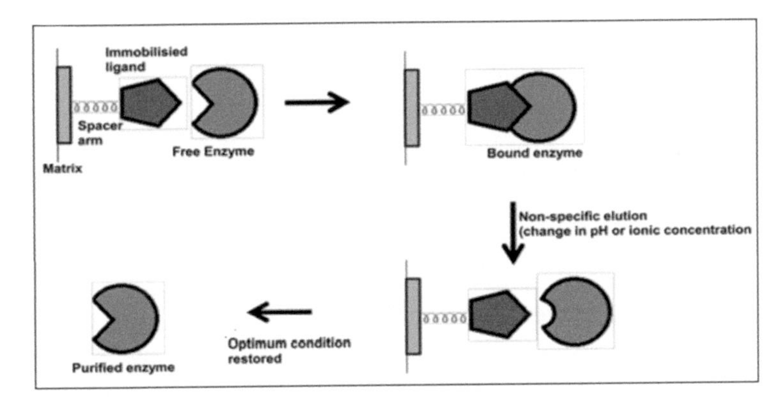

Figure 15.5 Affinity chromatography principle.

Affinity Chromatography Types

- Antibody–antigen, enzyme-inhibitor, and enzyme-substrate belong to the **bio-affinity chromatography**.
- The Cibacron Blue F3G-A dye coupled to the dextran matrix has a strong affinity toward dehydrogenases. This comes under the class of **dye affinity chromatography**.
- Fe^{2+}, Ni^{2+}, or Zn^{2+} is coupled to the matrix. The matrix-bound metal forms a multidentate complex with protein containing poly-histidine tags. This is an example of **metal affinity chromatography**.

Advantages:

- Affinity chromatography utilizes molecular size, charge, hydrophobic patches, and isoelectric points. Thus, it is more specific in detecting analytes in comparison to other purification techniques.
- In affinity purification, more than 90% recovery is possible.
- Affinity purification is reproducible and provides consistent results from one purification to another because it is independent of the presence of contaminating species.
- Affinity purification is extremely robust and easy to perform.

Gas Chromatography

- Gas chromatography is used for the separation of volatile substances such as methyl ethers of fatty acids.
- In this technique, a glass or metal column is filled with an inert solid such as silica gel coated with a nonvolatile liquid (e.g., silicone oil).
- The vaporized sample is introduced at one end of the column and is carried through the column by a stream of an inert gas such as nitrogen.
- Each component in the sample moves through the column, determined by its partition coefficient between the mobile phase and the stationary liquid.
- Individual components of the sample emerging from the column in the gas phase are detected by physical or chemical means.

- The chromatogram patterns consist of several peaks, each corresponding to one compound.
- The area under each peak is proportional to the concentration of the compound.
- The time taken to emerge from the column is characteristic of each compound.

Based on the force per unit area of pressure, liquid chromatography is often classified into the following categories:

- Low pressure liquid chromatography contains a pressure limit less than 5 bar.
- Medium pressure liquid chromatography contains a pressure limit of 6 bar to 50 bar.
- High pressure liquid chromatography contains a pressure limit from 50 bar to 350 bar.

High-Pressure Liquid Chromatography

- A typical polysaccharide bead is not appropriate to withstand the high pressure during high-pressure liquid chromatography (HPLC).
- HPLC silica-based beads are recommended due to high pressure, and the smaller size of the beads gives a higher number of theoretical plates.
- This provides superior resolution to HPLC for separating complex biological samples.
- A mixer is required to combine the buffer received from both pumps to make a linear or step gradient.
- The column is made of glass or steel. A UV-visible detector is used.
- The eluent can be collected in different fractions by a fraction collector.
- The profile of the eluent with respect to the measured property in a detector can be plotted in the recorder.

Applications:

- Analysis of antibiotics
- Detection of endogenous neuropeptides in brain extracellular fluids

Multiple Choice Questions

1. Which of the following techniques uses the difference in the net charges of proteins at a given pH?
   a. Thin layer chromatography
   b. Ion exchange chromatography
   c. High performance liquid chromatography
   d. Paper chromatography

Ans: b

2. The frequently employed materials for the adsorption chromatography of proteins include
   a. High-capacity supporting gel
   b. Starch blocks
   c. Calcium phosphate gel, alumina gel, and hydroxy apatite
   d. All of these

Ans: c

Chapter 16

Electrophoresis

Electrophoresis is the movement of charged particles through an electrolyte when exposed to an electric field. Electrophoresis literally means walking in an electric field. Negatively charged particles (anions) move to the anode, and positively charged particles (cations) move to the cathode.

The speed of movement of ions in the electric field depends on the following factors:

- Particle size and shape
- Net charge of the molecule
- Electric field strength
- Characteristics of carrier medium
- Operating temperature

Types of Electrophoresis

- Zone electrophoresis
- Gel electrophoresis
- Paper electrophoresis
- Cellulose acetate electrophoresis
- Thin layer electrophoresis
- Mobile boundary electrophoresis
- Isoelectric focusing
- Capillary electrophoresis
- Immunoelectrophoresis

Clinical Biochemistry: A Laboratory Guide
Rooma Devi, Aman Chauhan, Simmi Kharb, and Chandra Shekhar Pundir

ISBN 978-981-4968-75-1 (Hardcover), 978-1-003-45566-0 (eBook)
www.jennystanford.com

Paper electrophoresis

Serum proteins are separated using Whatman #1 filter paper, veronal buffer or tris buffer at pH 8.6, and stained Amido Black or Bromophenol Blue. A few microliters of a solution of charged particles (protein, DNA) is applied to the carrier medium in the form of a narrow band.

Gel electrophoresis

Polymer or acrylamide crosslinked with methylene bisacrylamide is used as a support medium. A thin slab of polyacrylamide, a glass plate, or a short column is used. The resolution is far better in this technique.

- The medium is placed in an electrophoretic chamber containing compartments on either side of the platform on which the support medium is placed. The compartments are filled with equal volumes of buffer solution of a suitable pH.
- The electrode is dipped into the buffer. A direct current is allowed to flow through.
- Based on charge and densities, proteins will move toward the cathode or the anode.
- The protein band is stained with a dye.

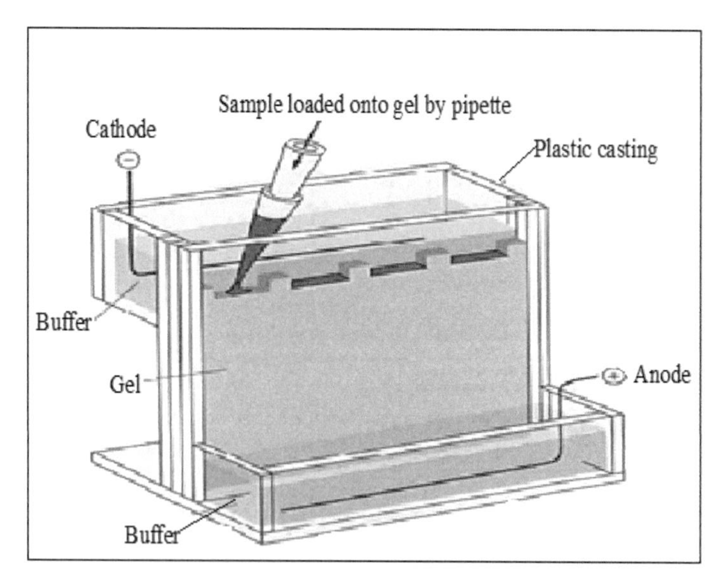

Figure 16.1 Diagram of gel electrophoresis.

Components of Electrophoresis

- Electric power (driving force)
- The support medium
- Buffer
- Sample
- Detecting system

The gel is mounted between two buffer chambers containing separate electrodes in most electrophoresis units so that the only electrical connection between the two chambers is through the gel.

Agarose and Polyacrylamide

- Both agarose and polyacrylamide are porous gels but have very different physical and chemical structures.
- Agarose is used to separate macromolecules such as nucleic acids, large proteins, and protein complexes.

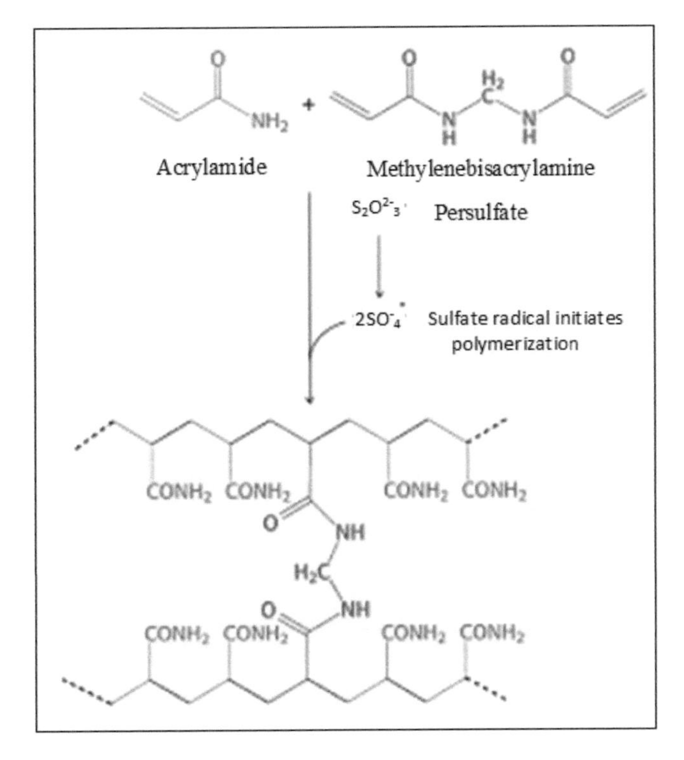

Figure 16.2 Crosslinking acrylamide chains.

- Polyacrylamide, which gives a small-pore gel, is used to separate most proteins from small oligonucleotides.
- Both are relatively electrically neutral.
- Polyacrylamide gels are stronger than agarose gels.
- The acrylamide monomer polymerizes to the long chains covalently bonded by the crosslinking agent.
- Polyacrylamide is chemically complex, so is the production and use of gels.

Sodium Dodecyl Sulphate (SDS)–PAGE

- A much easier method for estimating the molecular weight of proteins and protein subunits is electrophoresis on a polyacrylamide gel medium in the presence of sodium dodecyl sulfate (SDS).
- SDS interacts with the nonpolar interior of the protein molecule and unfolds the native structure.
- The charged SO_4^{2-} groups of the protein SDS complex contribute negative charge.
- Prior treatment of protein (containing disulfide bridges) with beta-mercaptoethanol is mandatory, as being reducing agent, it is responsible for protein unfolding and make its structure (structure of protein) into primary structure.
- The rate of mobility is inversely proportional to the size of protein.

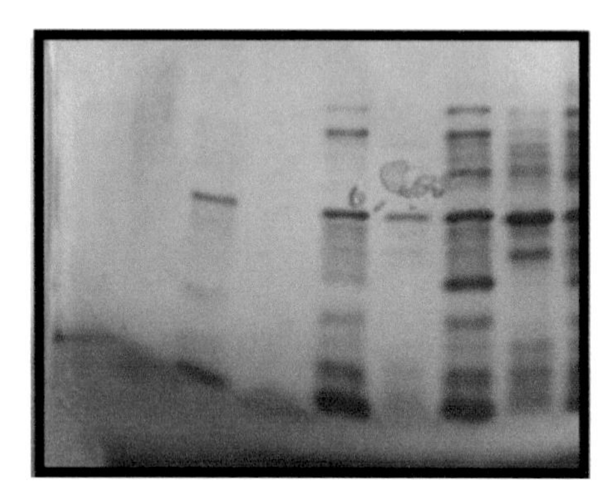

Figure 16.3 Coomassie Brilliant Blue staining of polyacrylamide gel.

- In SDS–PAGE, the protein SDS complex always moves toward the anode.
- To determine the molecular weight of an unknown protein, several standard proteins of known molecular weights are subjected to electrophoresis along with the unknown protein under identical conditions.
- Protein bands are stained with suitable dyes, such as Coomassie Brilliant Blue or silver stain, etc.

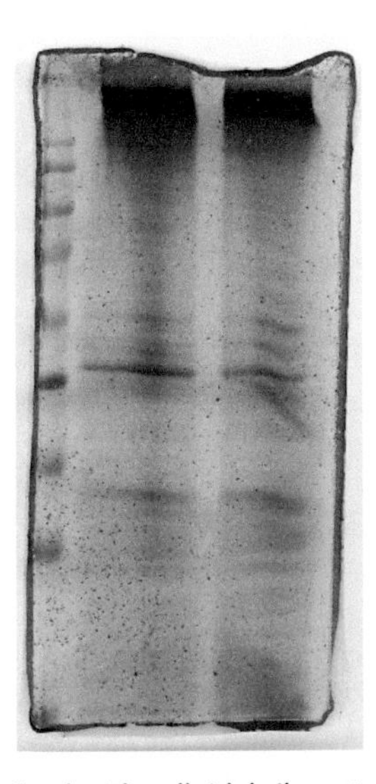

Figure 16.4 PAGE gel stained with colloidal silver stain.

Isoelectric Focusing

- This technique is primarily based on the immobilization of molecules at the isoelectric point pH during electrophoresis.
- This is the process by which proteins are separated according to the isoelectric point on a pH gradient.
- Focusing is done in two stages. First, a pH gradient is formed. In the second stage, the protein moves to the anode if the net charge is negative and to the cathode if the net charge is positive.
- When the protein reaches the isoelectric point (pI) on the pH gradient, the net charge becomes zero and migration stops.

- A stable pH gradient across the pH range is established (usually in the gel) to limit the electrical points of the components in the mixture.

Blotting Technique

- Blotting is used to switch proteins or nucleic acids from a slab gel to a membrane along with nitrocellulose, CM paper, nylon, or DEAE.
- The switch of the pattern may be executed with the aid of capillary or Southern blotting for nucleic acids or with the aid of electrophoresis for proteins or nucleic acids.

Immunoelectrophoresis

- This method is a mixture of both electrophoresis and immunological reactions.
- It is beneficial for the evaluation of a complicated combination of antigens and antibodies.

Applications

The following analytes can be separated by electrophoresis:

- Proteins and peptides
- Nucleotides and nucleosides
- Hemoglobin variants
- Lipoproteins
- Organic acids
- Isoenzymes

Multiple Choice Questions

1. For which of the following is electrophoresis **now no longer** used?
   a. Protein separation
   b. Lipids separation
   c. Nucleic acid separation
   d. Amino acids separation

Ans: b

2. Which of the following factors does not influence electrophoretic mobility?
   a. Shape of molecule
   b. Size of molecule
   c. Stereochemistry of the molecule
   d. Molecular weight

Ans: c

Practical Spots in Biochemistry

Spot 1

1. Identify the instrument.
2. Write a note on its principle and clinical applications.

Chapter 17

Enzyme-Linked Immunosorbent Assay

Enzyme-linked immunosorbent assay (ELISA) is an analytical method for detecting a substance through a specific interaction between an antibody and its antigen. This extremely sensitive technique is used for detecting and quantifying substances such as antigens, antibodies, hormones, and proteins. ELISA has four basic steps, which are summarized in Fig. 17.1.

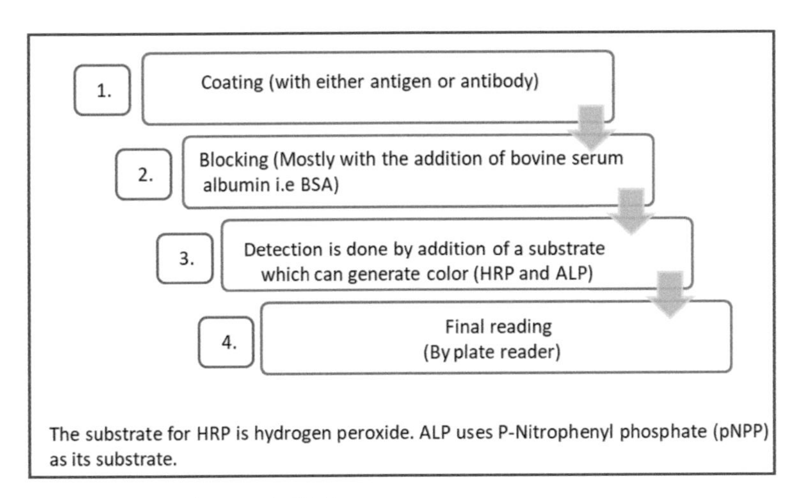

Figure 17.1 Basic steps in ELISA.

Clinical Biochemistry: A Laboratory Guide
Rooma Devi, Aman Chauhan, Simmi Kharb, and Chandra Shekhar Pundir

ISBN 978-981-4968-75-1 (Hardcover), 978-1-003-45566-0 (eBook)
www.jennystanford.com

Types of ELISA

Direct ELISA (Antigen-Coated Plate to Screen Antibodies)

In direct ELISA, the first step is immobilization of the antigen on the 96 wells ELISA plate, followed by incubation at 37°C for 1 h or 4°C overnight. Before incubation, the ELISA plate must be covered properly with a cover sheet provided by the manufacturer of the kit. After incubation, discard the coating solution and wash it three times by using wash buffer (to wash the plates of any potential unbound). Now block any remaining protein-binding sites in the coated wells (by 200 μl/well of blocking buffer) followed by covering, incubation, aspiration, and washing.

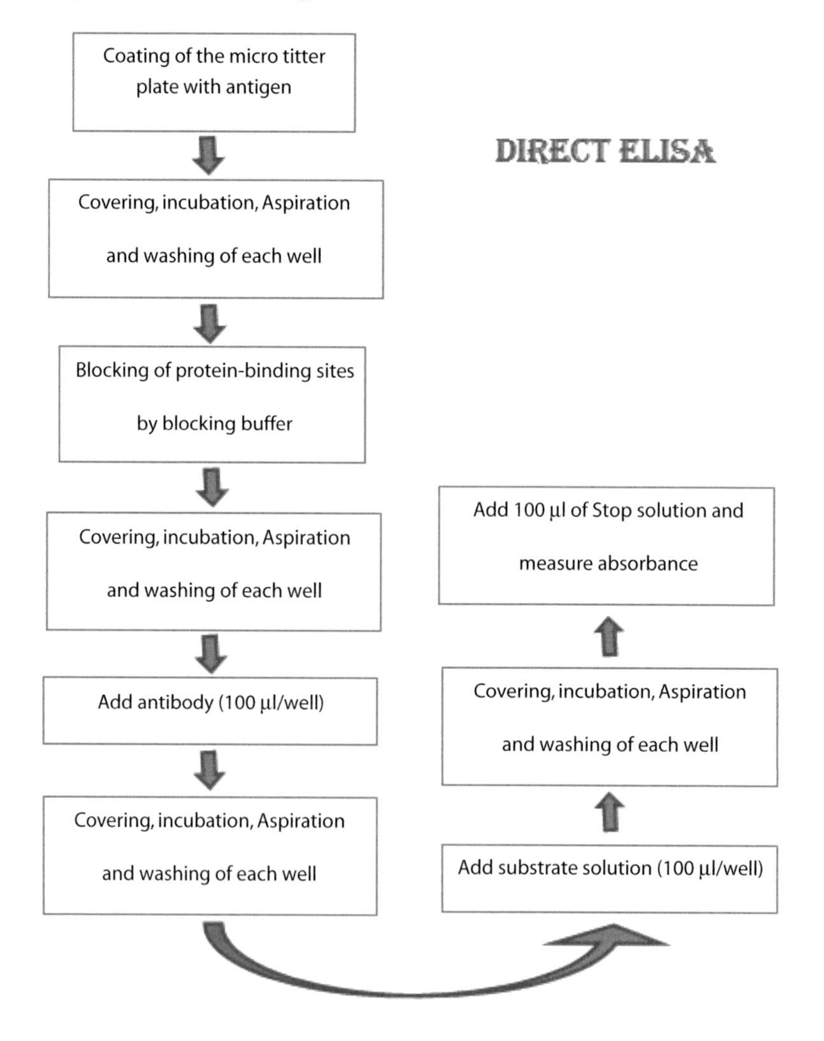

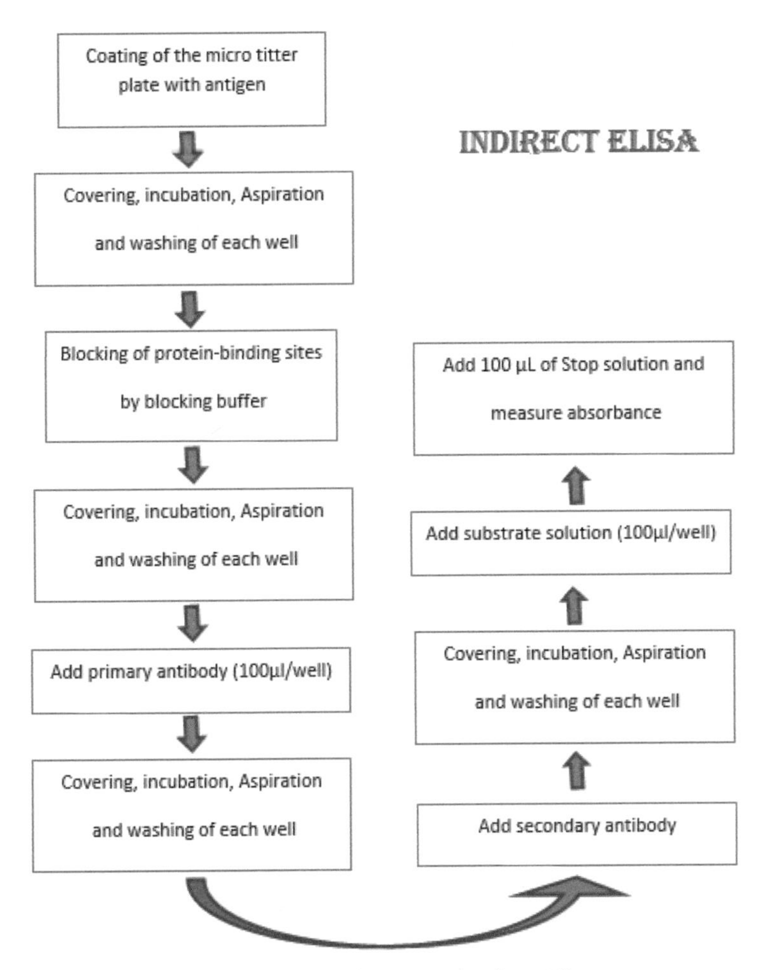

Figure 17.2 Schematic flowchart of Direct and Indirect Elisa.

Now add antibody (100 µl/well) diluted in blocking buffer followed by covering, incubation, aspiration, and washing. Now add the substrate solution (100 µl/well) and repeat all steps again (covering, incubation, aspiration, and washing) and now finally add 100 µl/well of stop solution.

Now read the absorbance (optical density) using a plate reader. The absorbance will be directly proportional to the concentration of the analyte.

Direct ELISA is a simple process but less specific than sandwich ELISA. **Indirect ELISA (Antigen-Coated Plate, to Screen Antigen/ Antibody)**

The basic steps of indirect ELISA are the same as those of direct ELISA. The first step is immobilization of the antigen on the 96 wells ELISA plate, which is followed by covering, incubation, aspiration, and washing. After incubation, discard the coating solution and wash it three times by using a wash buffer (to wash the plates of any potential unbound). Now block any remaining protein-binding sites in the coated wells (by 200 μl/well of blocking buffer) followed by covering and incubation at room temperature for 1–2 h after covering the plate properly. Wash the plate thrice using a wash buffer again.

Indirect ELISA requires two antibodies: a primary detection antibody that attaches to the protein of interest and a secondary enzyme-linked antibody complementary to the primary antibody. The first primary antibody will be added followed by covering, incubation, aspiration, and washing, and then the secondary enzyme-linked antibody is added and incubated. After this, the steps are the same as those of direct ELISA, which include a wash step, addition of substrate, and detection of a color change.

Indirect ELISA has a higher sensitivity when compared to direct ELISA [contains a polyclonal antibody (pAb), which recognizes the different epitopes of the primary antibody].

Sandwich ELISA (Antibody-Coated Plate; Screening Antigen)

Sandwich ELISA begins with a capture antibody coated onto the wells of the plate. It is termed a "sandwich" because the antigens get sandwiched between two layers of antibodies (capture and detection antibodies). Add capture antibodies to the plates followed by covering, incubation, aspiration, and washing. The next step is blocking the protein-binding site by blocking the buffer followed by covering, incubation, aspiration, and washing. Now pipette 100 μl of the sample into each well followed by covering, incubation, aspiration, and washing. Now add 100 μl of detection antibodies and repeat the basic steps again. Now add 100 μl of secondary antibodies followed by covering, incubation, aspiration, and washing. Now add the substrate solution (100 μl/well) and repeat all steps again (covering, incubation, aspiration, and washing). Now add 100 μl/well of the stop solution. Finally, measure the absorbance by the plate reader. This technique is highly sensitive and specific than direct ELISA. (Two antibodies are used to bind the protein of interest.)

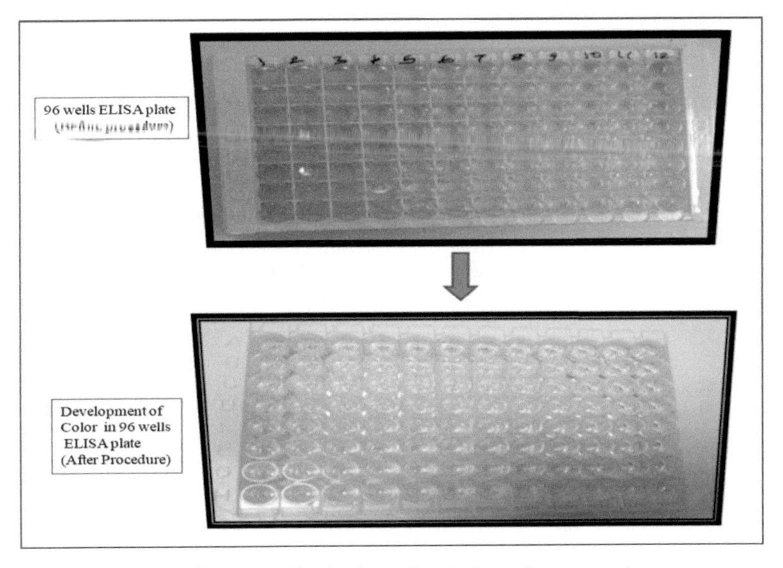

Figure 17.3 Development of color in wells of plate after procedure.

Competitive ELISA (Screening Antibody)

This type of ELISA utilizes two specific antibodies: an enzyme-conjugated antibody and another antibody present in the test serum (if the serum is positive). Combining the two antibodies into the wells will create a competition for binding to the antigen. A change in color means that the test is negative because the enzyme-conjugated antibody has bound the antigens (not the antibodies of the test serum). The absence of color indicates a positive test and the presence of antibodies in the test serum. Competitive ELISA has a low specificity and cannot be used in dilute samples. However, its benefit is that less sample purification is needed.

Detection is carried out by the addition of a substrate that can generate a color. Many substrates are used in ELISA. However, the most commonly used substrates are horseradish peroxidase (HRP) and alkaline phosphatase (ALP). The substrate for HRP is hydrogen peroxide, which results in a blue color. ALP measures the yellow color of nitrophenol after incubation at room temperature for 15–30 min and usually uses P-nitrophenyl phosphate (pNPP) as its substrate.

SANDWICH ELISA

Coating of the micro titter plate with capture antibody

Covering, incubation, Aspiration and washing of each well

Blocking of protein-binding sites by blocking buffer

Covering, incubation, Aspiration and washing of each well

Pipette 100 μL of samples into each well

Covering, incubation, Aspiration and washing of each well

Pippete100 μL of the detection antibody solution into each well

Covering, incubation, Aspiration and washing of each well

Pippete100 μL of the secondary antibody into each well

Covering, incubation, Aspiration and washing of each well

pipette 100 μL of substrate in each well

Covering, incubation, Aspiration and washing of each well

Add 100 μL of Stop solution and measure absorbance

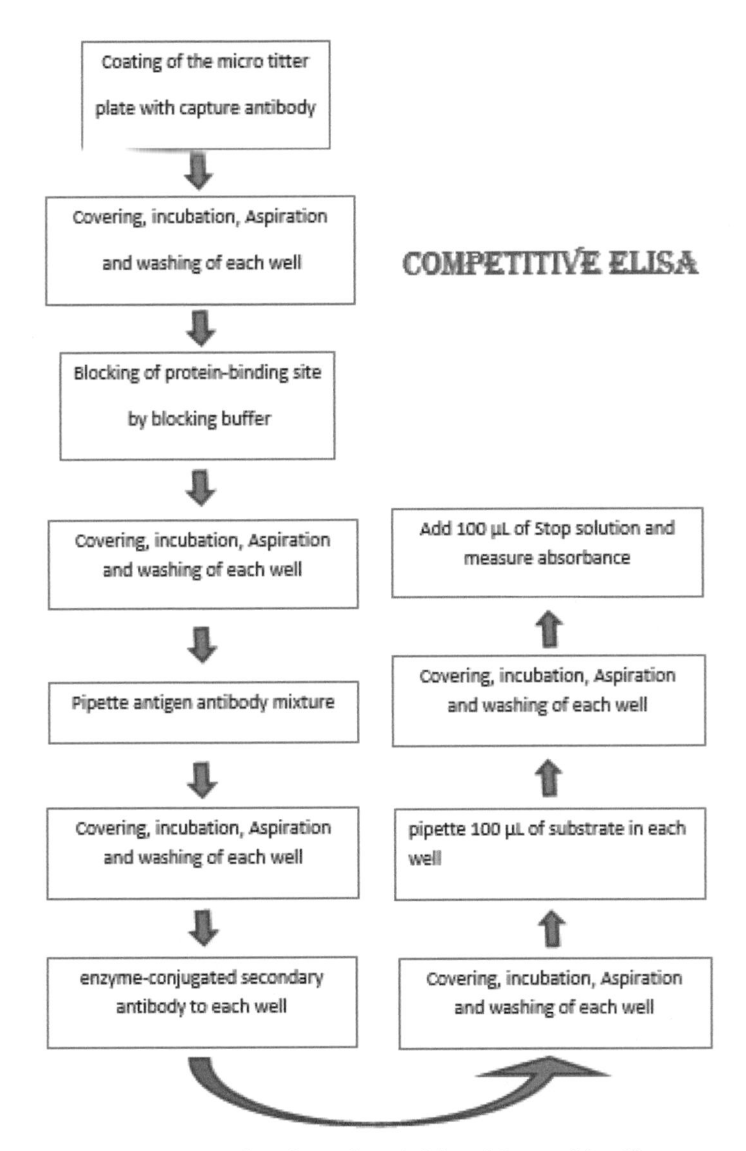

Figure 17.4 Schematic flowchart of Sandwich and Competitive Elisa.

Among the above four steps, the mandatory step is "washing" of the plate using a buffer [by phosphate-buffered saline (PBS) and a non-ionic detergent, to remove the unbound material]. It has the great advantage of quantifying small molecules.

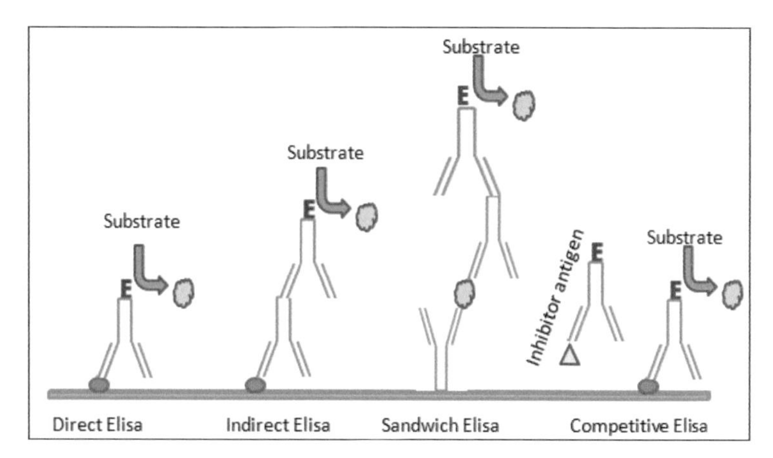

Figure 17.5 Schematic flowchart of various types of ELISA.

Although ELISA is a simple, highly sensitivity, and specific process, it also has some disadvantages, such as refrigerated storage and transport of antibodies, expensive, instability of antibodies, etc.

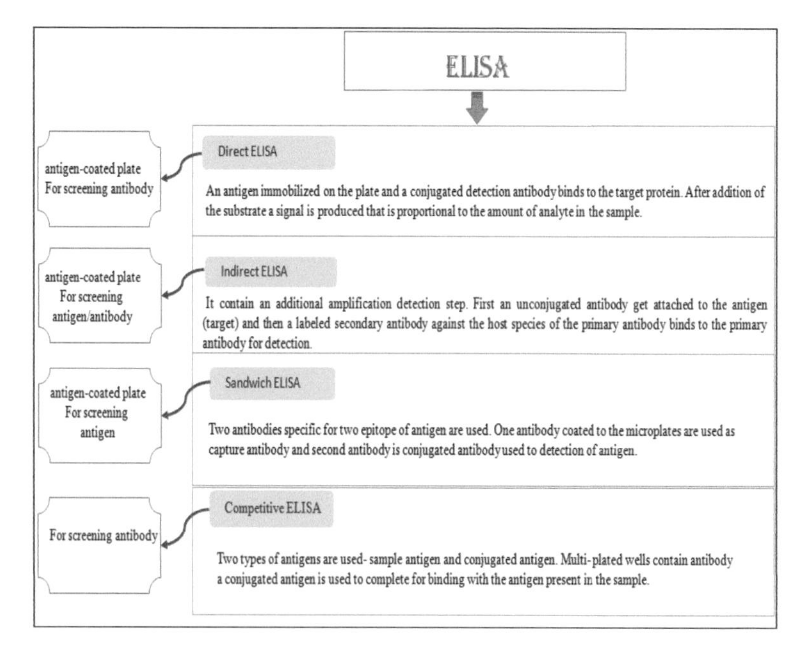

Applications of ELISA

- To detect food allergens in the food industry

- To detect drugs abuse (amphetamine, cocaine, etc.)
- To estimate tumors markers (PSA, CEA, etc.)
- To estimate and detect hormones levels (prolactin, testosterone)
- To estimate and detect antibodies in blood (anti-dsDNA, ANA, HIV, hepatitis)

Multiple Choice Questions

1. Which of the following principles is used in ELISA?
   a. Antigen–antibody interaction
   b. Detection based on amplification
   c. Both of the above
   d. None of the above

Ans: a
The detection is based on antigen–antibody interactions in all types of ELISA.

2. How many antibodies are used in sandwich ELISA?
   a. 1
   b. 2
   c. 3
   d. 4

Ans: b
In sandwich ELISA, two types of antibodies are used: primary (coated to the microplate) and secondary (conjugate antibody to detect antigen).

3. In which of the following, only one antibody is used?
   a. Direct ELISA
   b. Indirect ELISA
   c. Sandwich ELISA
   d. All of the above

Ans: a
In indirect and sandwich ELISA, two antibodies are used. Direct ELISA is a simple process, which uses only one antibody (less specific than sandwich ELISA).

4. ELISA can be used for the detection/estimation of:
   a. PSA

b. Cocaine
c. HIV
d. All of the above

Ans: d

ELISA is a technique based on the principle of antigen–antibody interaction. It is used in hormonal assay, drug abuse, detection of food allergens, detection of HIV and hepatitis, etc.

5. Probes are used in all of the following techniques except:
   a. FISH
   b. ELISA
   c. PCR
   d. None of the above

Ans: b

ELISA is an analytical technique based on the principle of antigen–antibody interactions. In FISH (fluorescent in situ hybridization) and PCR (polymerase chain reaction), chromosomes are exposed to a small DNA sequence called probe.

Chapter 18

Radioimmunoassay

Radioimmunoassay (RIA) is a method developed by Rosalyn Yalow and Solomon Berson in 1959 to measure insulin in plasma. In 1977, Yalow won the Nobel Prize for the development of RIA. It is also known as competitive radioassay. Being a very sensitive technique, it can detect the concentration of analytes in nanograms (0.001 µg/mL). It is an important technique in clinical biochemistry for the estimation of hormones, steroids, and drugs, which are present in minute quantities and cannot be detected by the colorimetric or other chemical methods. It is used for measuring the antigen concentration of a patient's sample. RIA is a combined technology of nuclear medicine (tracer technique) and immunology.

Principle

- RIA is a competitive assay, where a competition between unlabeled and labeled antigen is seen to quantify and detect an antigen. In this assay, a fixed amount of antibody is used, then sample containing antigen (unlabeled antigen) are allowed to react with the constant and fixed amount of antibody which is already bound with radio-labeled antigen. The labeled antigen will get displaced by the unlabeled antigen from sample (if sample contains more antigens). The amount of radio-labeled antigen displaced is directly proportional to the amount of unlabeled antigen in the sample. It can measure the analytes

Clinical Biochemistry: A Laboratory Guide
Rooma Devi, Aman Chauhan, Simmi Kharb, and Chandra Shekhar Pundir

ISBN 978-981-4968-75-1 (Hardcover), 978-1-003-45566-0 (eBook)
www.jennystanford.com

present in very low concentration, so it is a very sensitive and specific method for detection.

- RIA requires a sample containing a specific antigen (T_3 or T_4) of interest, a complementary antibody, and a radiolabeled version of the antigen (T_3 or T_4 having I^{131}).
- Incubate the sample radiolabeled antigen (T_3 or T_4 having I^{131}) and antibody and form an antigen–antibody (Ag–Ab) complex.
- This Ag–Ab complex is then treated with the patient's sample having an unlabeled antigen.
- The unlabeled antigen displaces the labeled antigen, which is then separated and the radioactivity per minute is calculated. This is used to calculate the unlabeled antigen in the patient's blood (Fig. 18.1).

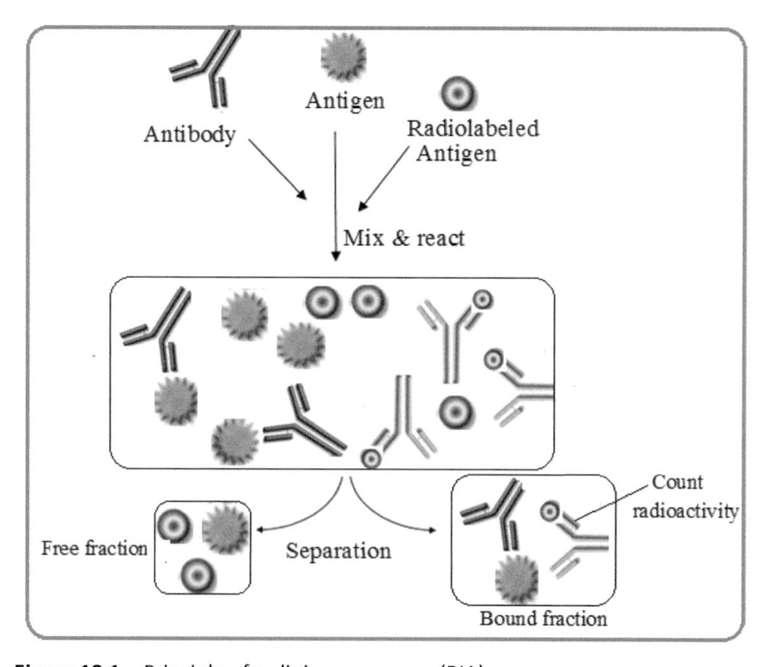

Figure 18.1 Principle of radioimmunoassay (RIA).

Reagents Required for RIA

- **Labeled antigen (tracer):** High-purity Ag should be used. Pure Ag is labeled with a radioisotope. The protein antigen tyrosine is usually labeled I^{125}. Non-protein antigens are labeled with tritium.

- A binder (antibody) that is a specific antiserum
- A separation system for separating the "bound" and "free" phases.
- Standard antigen (unlabeled antigen of known concentration) must be in pure form.
- Free human antiserum.
- Used radioisotope beta emitters 3H and 14C, gamma emitter 125I.
- Appropriate pH 7–9 buffer.
- **Scintillation counter:** Gamma counter for I125-labeled antigen. This marker has a very high sensitivity and a low half-life. Beta counter for H3-labeled antigen. It is less sensitive but has a longer half-life.

Procedure:

- A series of test tubes are taken in which varying amounts of pure antigen are incubated with fixed amounts of the labeled antigen and antibody for 6 h.
- The free and the bound antigens are separated, and the radioactivity in the bound Ag*Ab fraction is calculated.
- The test sample with the unknown amount of antigen is processed similarly.
- Standards are used to plot the standard curve.
- By referring to the standard curve, the amount of antigen can be obtained.

Advantages of RIA

- RIA can be used to assay any compound, which is immunogenic, available in pure form, and can be radiolabeled.
- It can be used to assay many non-protein compounds also, for which antibodies can be produced, e.g., hormones and drugs (mainly insulin and other hormones).
- It is a very sensitive, highly specific, and reproducible method.
- It has a broad range of measurements.

Disadvantages of RIA

- High cost of equipment and reagents
- Risk of radiation damage
- Cumbersome procedure
- Limited assay range
- Half-life of I^{125} is 60 days, which requires frequent ordering.
- Risk of radiation hazards to health
- Requirement of trained personnel
- Lengthy procedure
- Availability of excellent antisera

Applications of RIA

- Used in clinical chemistry to measure peptides, steroid hormones such as T_3, T_4, TSH, hCG, progesterone, etc.
- Drug detection
- Evaluation of methods for measuring plasma oxytocin
- Measurement of plasma estrogen

Chapter 19

Polymerase Chain Reaction

Amplification techniques can be classified into the following:

- **Thermal cycling:** Temperature of the reaction varies [polymerase chain reaction (PCR) and ligase chain reaction (LCR)].
- **Isothermal cycling:** Temperature of the reaction remains constant [nucleic acid sequence-based amplification (NASBA) and branched DNA technology].

Karry B. Mullis and coworkers developed PCR in the early 1980s. It is a laboratory technique for amplifying specific segments of DNA. It is the exponential amplification of a sample. At the end of a PCR reaction, billions of copies of the specific sequence are accumulated. The products of PCR are called amplicons.

Pre-requisites for PCR

- DNA sample (to be amplified)
- Deoxynucleotides (dNTPs)
- Thermostable DNA polymerase (resistant to heating and cooling cycles in PCR)
  a. Taq polymerase: grow in hot spring
  b. Pfu
  c. Pwo

Clinical Biochemistry: A Laboratory Guide
Rooma Devi, Aman Chauhan, Simmi Kharb, and Chandra Shekhar Pundir

ISBN 978-981-4968-75-1 (Hardcover), 978-1-003-45566-0 (eBook)
www.jennystanford.com

- $MgCl_2$ and KCL (divalent cations such as Mg^{2+} and Mn^{2+} stabilize the buffer solution)

If the incorporation of dNTP is not accurate, Taq polymerase cannot correct it as it does not have proofreading property. Some polymerases bearing this property are Pfu and Pwo.

Table 19.1 DNA polymerase used in PCR

DNA Polymerase	Source	Proofreading Activity
Taq polymerase	*Thermus aquaticus*	No
Pfu	*Pyrococcus furiosus*	Yes
Pwo	*Pyrococcus woesei*	Yes

Steps of PCR Cycle

PCR involves a series of cycles of three successive reactions. The product of the DNA acts as a template strand for further DNA synthesis.

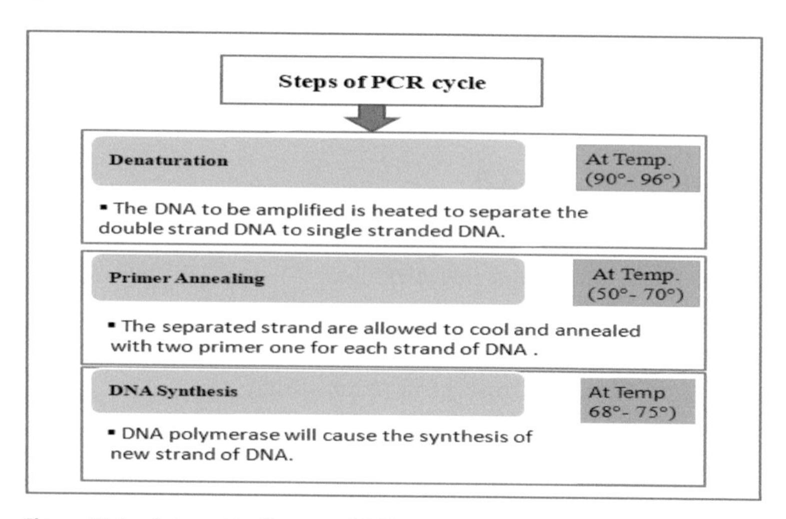

Figure 19.1 Schematic diagram of PCR.

Primer

Primers are short nucleotide sequences, which can bind to the target DNA. So prior knowledge of the DNA sequence (to be amplified) is mandatory. A DNA polymerase can add a nucleotide only onto a pre-existing 3'-OH group; it needs a primer to which it can add the first nucleotide.

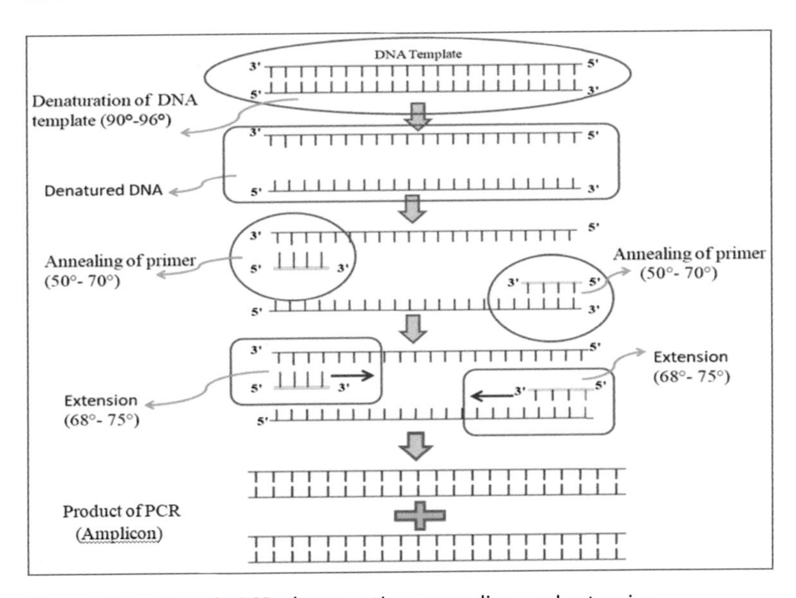

Figure 19.2 Steps in PCR: denaturation, annealing, and extension.

Types of PCR

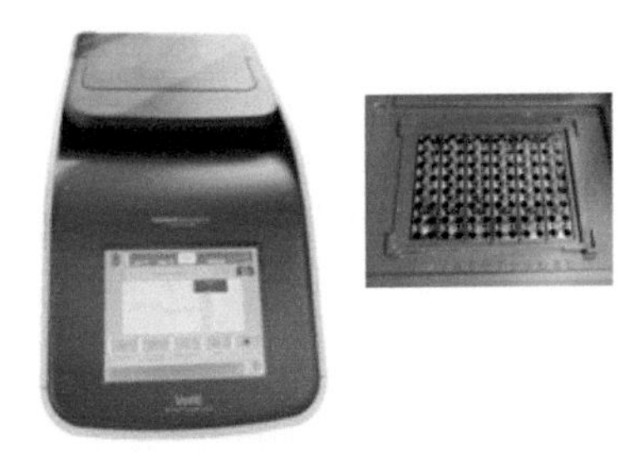

Figure 19.3 PCR machine (thermal cycler, DNA amplifier) used to amplify the segment of DNA.

Table 19.2 Different types of PCR used for amplification of specific segment

	PCR Type	
1.	Conventional PCR	Amplification by use of two primers.
2.	Qualitative real-time PCR (qPCR)	It allows the detection of dsDNA during amplification. qPCR thermocyclers are equipped with the ability to excite fluorophores at specific wavelengths, detect their emission with a photodetector, and record the values. The fluorophores used in qPCR are ethidium bromide, SYBR green, or sequence-specific probe (TaqMan probe).
3.	RT-PCR (reverse transcription PCR)	It is carried out by reverse transcriptase, which converts RNA (mRNA) to complementary DNA (cDNA), which works as a template for new DNA synthesis.
4.	Nested PCR	Two sets of primers are used in this type of PCR.
5.	Arbitrary PCR	Random primers are used complementary to the template DNA.
6.	Inverse PCR	It is used to amplify the flanking region of DNA.

Applications of PCR

- This test is very efficient in detecting viruses having long latency periods, such as HIV. It is a very sensitive method for detecting viral RNA/DNA sequences. PCR tests are used to detect the genetic material of HIV, called RNA. It can detect very early infections before antibodies have been developed and may be performed just days or weeks after exposure to HIV virus. PCR is not performed as often as other HIV tests (although it is more accurate) because of its cost on individual samples.
- PCR is a sensitive technique for the quantification of both nuDNA (nuclear DNA) and mtDNA (mitochondrial DNA). PCR methods based on mitochondrial genes have been used in **forensic science** because of their high copy number per cell and lack of recombination. DNA profiling (DNA typing, genetic fingerprinting, DNA testing) is a technique used in forensic

science to identify criminals based on their DNA profile. Tiny samples of DNA isolated from a crime scene may be compared with the DNA from crime suspects or a DNA database. PCR-based DNA fingerprinting can also be used in paternity disputes in which the DNA of an individual is compared with their close relatives.

- Although most methods of **mutation detection** depend on PCR, PCR itself does not detect the mutation. Rather, the products of PCR (amplicons) are further analyzed by some other method to find possible mutations, e.g., conformation-based techniques such as single-stranded conformational polymorphism (SSCP) analysis, denaturing gradient gel electrophoresis (DGGE), or sequencing. Some methods of PCR do not need further analysis and can detect mutation directly, such as real-time PCR, the amplification refractory mutation system (ARMS), and quantitative fluorescent PCR (QF-PCR).
- PCR is very useful in the detection of **viral load** for the diagnosis of a particular disease, e.g., COVID-19. The cycle threshold (Ct) is defined as the number of cycles required for the fluorescent signal to cross the threshold. Ct levels are inversely proportional to the amount of target nucleic acid in the sample. The lower the Ct level, the greater the amount of target nucleic acid in the sample.
- Nowadays, rapid and low-cost methods of diagnosis are becoming important for detecting **single-gene disorders and chromosomal abnormalities**. PCR does not allow the diagnosis of all congenital defects; only a limited number of diseases can be diagnosed by PCR, such as cystic fibrosis.
- PCR is widely used for the DNA analysis of **archaeological specimens**. PCR can be used in the genetic analysis of human, animal, and plant specimens. Ancient DNA can be extracted from historical objects such as parchments or leathers, or from plant and animal remains (bones, teeth, and hair) and can be analyzed further by PCR.
- PCR is also used to detect length polymorphism such as VNTR and gene expression and to establish precise tissue types for transplantation.

Multiple Choice Questions

1. Karry B. Mullis developed the following technique:
   a. Mass spectrometry
   b. PCR
   c. Flow cytometry
   d. None of the above

Ans: b

2. PCR is used for
   a. Sequencing
   b. DNA degradation technique
   c. Amplification of DNA
   d. None of the above

Ans: c

Chapter 20

Biosensors/Glucometer

Biosensors are analytical tools consisting of a substrate or an analyte and a specific interface in close proximity to or incorporated with a transducer to permit the quantitative development of some complex biochemical parameters. Thus, both the substrates/analytes and the transducers are significant parts of these insightful devices, which consist of an immobilized biorecognition component (or bioreceptor) such as proteins (e.g., cell receptors, catalysts, and antibodies), nucleic acids, microorganisms, or even entire tissues that interact with specific types of analytes. The signals produced due to the reactions between a biorecognition component and its specific target analyte(s) are converted into a detectable electrical or other signal by the transducers through a process known as signalization. The signals produced (electrical or optical) are usually proportionate to the number of analyte–bioreceptor interactions whose intensity may be directly or inversely proportional to the concentration of analytes. This chapter covers the different sorts of transducers that are utilized in various biosensors, their characteristics, a couple of important applications, and also new trends in transducers found in biosensors.

Clinical Biochemistry: A Laboratory Guide
Rooma Devi, Aman Chauhan, Simmi Kharb, and Chandra Shekhar Pundir

ISBN 978-981-4968-75-1 (Hardcover), 978-1-003-45566-0 (eBook)
www.jennystanford.com

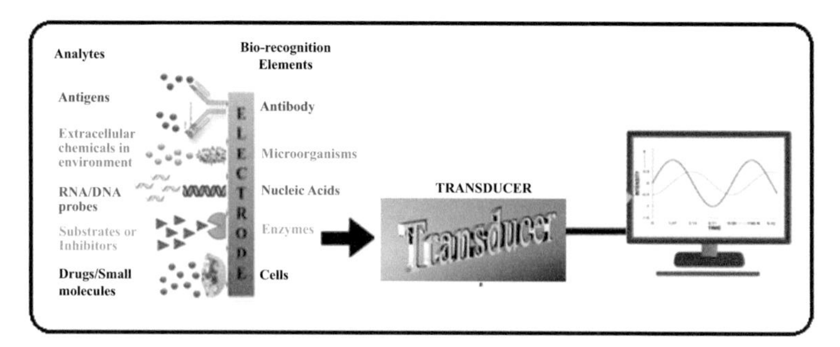

Figure 20.1 Structure and components of biosensors.

Transducers in Biosensors

To develop biosensors, suitable transducers have to be selected because a transducer converts the signal produced (biochemical) into a measurable signal, which plays an important role in the detection process of biosensors. Based on the transducing mechanism used in biosensors, they can be classified into the following types:

- Electrochemical transducers
- Optical transducers
- Calorimetric (thermometric) transducers
- Piezoelectric transducers
- Magnetic transducers

Some of these transducers can be further divided into various types, which will be discussed below.

Electrochemical

Transducers depending on electrochemical detection mechanisms are most commonly used in biosensors. The first electrochemical biosensor was developed by using glucose oxidase (GOx) for the detection of glucose by Clark and Lyons in 1962. After that many improvements have been made and new types of transducers have been developed.

Electrochemical transducers detect various kinds of electrochemical species, either generated or consumed, during the interaction of the biosensing element (enzyme/antibody) with the analyte (or substrate). Based on the mechanism used for the detection of electrochemical species, electrochemical transducers used in biosensors are mainly classified as (i) amperometric, (ii) potentiometric, (iii) conductometric, and (iv) photoelectrochemical.

- **Amperometric** transducers measure the movement of electrons produced in an oxidation–reduction reaction.
- **Potentiometric** transducers measure the change in electric potential caused by the distribution of charges.
- **Thermometric/calorimetric** transducers measure the heat output of the reaction.
- **Optical** transducers measure the light produced during the reaction or difference between the light absorbed by the reactants and products.
- **Piezoelectric** transducers measure the changes due to the mass of reactants and products.
- **Magnetic** transducers measure the variations in the magnetic properties or effects induced magnetically.

Amperometric

These transducers measure the electric current generated by various electroactive species during a biochemical reaction. The concentration of the analyte to be analyzed is linearly dependent on the amount of current produced usually in redox reactions. The amount of the current during the reaction changes due to an increase or decrease in the thickness of the diffusion layer at the electrode.

Chronoamperometry is another type of amperometric detection used in biosensors. In this technique, a steady-state current is measured as a function of time applying a square-wave potential to the working electrode.

Potentiometric

These transducers usually measure the potential difference between the two electrodes—reference electrode and working electrode—generated due to the redox reactions occurring between the bioreceptors and analytes at their surfaces. Any change in the concentration or activity of a specific analyte present in the solution leads to a change in potential. The relationship between the concentration of the analyte and the potential difference generated is governed by the Nernst equation:

$$EMF \text{ or } E_{\text{cell}} = E^0_{\text{cell}} - \frac{RT}{nF} \ln Q \tag{20.1}$$

where

- E_{cell} is the observed electrode potential at zero current, which is usually referred to as the electromotive force (EMF),
- E^0_{cell} is a constant potential contribution to the electrode,
- R is the universal gas constant,
- T is the absolute temperature in degrees Kelvin,
- n is the charge number of the electrode reaction,
- F is the Faraday constant, and
- Q is the ratio of the ion concentration at the anode to the ion concentration at the cathode.

Potentiometric measurement can be of two types: direct potentiometry and potentiometric titration.

1. **Direct potentiometry:** In this type, the concentration of analytes can be measured directly as per the Nernst equation. Ion specific electrodes (ISE) have been used in these potentiometric devices, which can detect only specific ions produced or consumed during the reaction even in very small amounts.
2. **Potentiometric titration:** It is another method for electrically detecting the endpoint in a biochemical reaction at which equal amounts of different solutions reach a state of equilibrium (e.g., 0.1 M HCl and 0.1 M NaOH). In this type of measurement, titration is performed and the endpoint is determined by the variations in the electrode potential, which occur during the changes in the concentration of the particular ion(s) present in the solution at constant or zero current.

Conductometric

Conductometric analysis is widely used for chemical systems in which many chemical reactions have occurred to produce or consume ionic species, which change the overall electrical conductivity of the solution. These transducers measure the change in electrical conductivity, which occurs in the solution between a pair of metal electrodes during the reaction between the analyte and the biosensing element. These transducers are frequently used to study the enzyme reactions where the reaction between the analytes and the immobilized enzymes, producing or consuming charged species, leads to a change in the ionic strength of the medium, which affects the electrical conductivity between the two electrodes.

Photoelectrochemical

Photoelectrochemical enzyme-based biosensors are a new subclass of biosensors that combine the selectivity of enzymes and the inherent sensitivities of bioanalysis. In a typical enzymatic biosensor, the enzymatic system upon irradiation transforms the specific biocatalytic events into electrical signals through interactions between the reaction chain catalyzed by enzymes and the semiconductor species.

Optical

These biosensors (or transducers) are used for detecting an optical signal generated during a biological reaction and/or any other chemical reactions. Optical biosensor measured the light produced during the reaction or difference between the light absorbed by reactants and products. Such biosensor can be used for measurement of pH O_2 and CO_2 which can generate an optical signal that is related to the concentration of target species in the sample.

Fluorescence

Fluorescence is a phenomenon in which a molecule absorbs light at a shorter wavelength (or longer frequency) and emits the light at a longer wavelength (or shorter frequency) in the visible regions of the electromagnetic spectrum. It is a broadly used optical detection-based method used for sensing the analytes that can show fluorescence. Optical detectors in fluorescence-based biosensors detect the change in the frequency of electromagnetic radiation between the excited and the emitted radiation during the reaction of analyte with the bio-recognition element. Fluorescence-based biosensors are very useful for biosensing in many analytes due to their selectivity and sensitivity.

Chemiluminescence

During some chemical reactions, different reactants react with each other and undergo an excited state. While forming the product, they again come back to the ground state along with the emission of light (luminescence). This process is known as chemiluminescence, as shown in the reaction given below. In this reaction, luminol reacts with hydrogen peroxide and produces 3-aminophthalate (3-APA) as a product along with the emission of light.

$$[\text{Luminol}] + [H_2O_2] \rightarrow [\text{3-APA}] + \text{Light}$$

Calorimetric (Thermometric)

Calorimetric biosensors are designed by attaching the biological component of a heat-sensing transducer known as **thermistor**. During a biochemical reaction, heat is either absorbed or produced, which induces a change in the temperature of the solution/medium. The construction of these biosensors consists of miniaturized thin-film thermistors and immobilized biological elements. Calorimetric transducers detect the changes in temperature that occur during the reaction between the biological element and its target analyte.

Piezoelectric

Piezoelectric biosensors are analytical devices that sense the affinity interaction between analytes and biorecognition elements, and they utilize crystals that undergo an elastic deformation when an electric potential is applied to them. Piezoelectric crystals (e.g., quartz) vibrate under the influence of an electric field. The frequency of this oscillation (f) depends on crystal's thickness and cut; each crystal has a characteristic resonant frequency. This resonant frequency changes as molecules adsorb or desorb from the surface of the crystal, obeying the relationship

$$\Delta f = Kf^2 \Delta m / A$$

where Δf is the change in resonant frequency (Hz), Δm is the change in mass of adsorbed material (g), K is a constant for the particular crystal which depends on factors such as its density, and A is the adsorbing surface area (cm^2). For any piezoelectric crystal, the change in frequency is proportional to the mass of absorbed material, up to about a 2% change. This frequency change is easily detected by relatively unsophisticated electronic circuits.

These changes occurring on the surface of the piezoelectric platform transduce into changes in piezoelectric effect, i.e., changes in the frequency of oscillation or surface acoustic waves, which are detected by the detector, and this frequency is mainly dependent on the elastic properties of the crystal.

Piezoelectric sensors can be further divided into two types on the basis of frequency detections: (i) bulk wave and (ii) surface acoustic wave (SAW). The sensors based on bulk waves can detect fundamental oscillations at higher frequencies, whereas SAW-based devices are capable of lower detection limits.

Magnetic

Due to the unique properties of magnetic materials, they are now being used in the development of biosensors to monitor biological interactions for quick detection at the point of test in different fields. In the past few years, magnetic nanoparticles (MNPs) have been developed as labels for biosensing technology for the detection, identification, localization, and manipulation of an extensive range of biological, physical, and chemical agents.

Applications

A variety of transducers are available for detecting diverse types of analytes, such as electrochemical, optical, electrical, and thermal transducers. Transducers are used in different types of biosensors, which are used in healthcare, food processing/monitoring, fermentation processes, biodefense, and many more. The main goal of researchers is using biosensors in the medical field. Different transducers used in biosensors can solve the problem of finding pathogens at very low concentrations and even at early stages of infection. Electrochemical detection is a very common method employed in biosensors for ultrasensitive detection of analytes. The electroactive species can exist as a labelling agent of an antibody/antigen or working solution or the generated electroactive species. Hence, electrochemical transducers are most beneficial mainly for onsite diagnostic tests as POC devices, which are accessible to doctors and patients. Moreover, biosensors based on these transducers have many benefits; they are usually portable, easy to handle, and user friendly. Other advantages include high sensitivity and specificity with real-time analysis, cost effectiveness, and easy to be miniaturized to hand-size devices. The glucose sensor is one of the most universal examples of enzyme-based biosensors, which can save millions of lives from diabetes. Colorimetric transducers are particularly attractive, as they can rapidly detect a disease, whereas an observation with the naked eye can only interpret a screening result. These kinds of transducers are very useful for qualitative detection of pathogens. The primary mechanism through which colorimetric transducers measure is the aggregation of nanoparticles, such as silver (AgNPs) and gold nanoparticles (AuNPs), which are distinguished by their size and ability to change colour significantly from red to blue when induced to aggregate and enhance the signal. Biomarkers are identified using the colorimetric assay in medical diagnostics.

Conclusion and Future Prospects

This chapter describes and characterizes different types of transducers used in construction biosensors. A biosensor device consists of a biosensing element intimately coupled or integrated within a transducer. Biosensors offer an exciting alternative to traditional analytical methods, which are applicable in various areas such as clinical diagnosis, food industry, environment monitoring, etc. and allow rapid, real-time, and multiple analyses simultaneously. Presently, the research on biosensor technology mainly emphasizes on the generation of more sensing biorecognition elements and transducers. Thus, biosensor technology offers an opportunity for the development of robust, low-cost, more sensitive, and accurate sensors for the detection of specific analytes. The next generation of biosensors based on nanostructures could lead to a construction of devices able to markedly compete with other analytical methods used today. The future depends on the development of new sensing elements and transducers.

Glucometer

A glucometer is a medical device for determining the approximate concentration of glucose in the blood. It is a portable instrument for the rapid measurement of blood glucose. Strips impregnated with glucose oxidase are allowed to react for a minute with a drop of blood obtained by fingertip puncture. The blood is then blotted, and the color developed in the strip is read with the instrument, which provides immediate result. This monitoring procedure can be performed at home or during a visit to a doctor's office. However, many people with diabetes are highly active and prefer to monitor their blood glucose concentration irrespective of where they are located. Therefore, lightweight, robust, and portable glucometers have substantial use.

Portable glucometers can be based on a variety of analytic techniques, including electrochemical or spectroscopic techniques. These glucometers generally display the results of blood glucose analysis on an LCD monitor or some other kind of local read-out mechanism, which can be read by users. This read-out mechanism need not be local, and there are advantages of using remote read-out devices via wireless devices, such as mobile phones. Wireless devices for analyzing blood glucose concentration allow the transmission of data to healthcare professionals who may collect the data for diagnostic purposes.

Types of Glucometers Based on Electrochemical Measurement

Approaches for designing glucose-monitoring systems are as follows:

- Amperometric
- Colorimetric

Glucometer Sensors (Test Strips)

These sensors use an electroenzymatic approach. That is, they use glucose oxidation with a glucose oxidase enzyme. The presence of glucose oxidase catalyzes the chemical reaction between glucose and oxygen, and the oxidation of glucose to gluconic acid results in an increase in pH, a decrease in oxygen partial pressure, and an increase in hydrogen peroxide.

The test strips measure changes in one or more of these components to determine glucose levels. The strips used in this design have three terminals or electrodes.

1. Reference electrode

2. Working electrode
3. Counter/auxiliary electrode

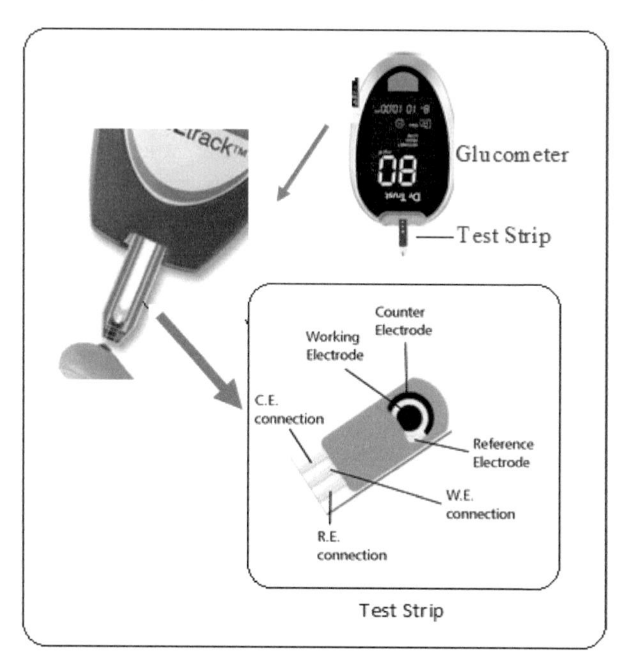

Figure 20.2 Diagram of blood glucose meter with test strips.

Advantages

- Reliable
- Provides high precision
- Can be used by patients without external help

Disadvantages

- An invasive method and requires precautions
- Expensive
- Test strips used have a short storage life
- Results vary due to changes in the operating environment
- The colorimeter used for analysis must be calibrated each time

Chapter 21

Glycated Hemoglobin

Glycated hemoglobin (HbA1c) is an additional marker besides the standard glucose and glycemia analyses, which has become a relevant marker in new analytical methods. As discussed in the following paragraphs, the determination of HbA1c is substantial for the diagnosis of diabetes and provides substantial results compared to the simple measurement of glycemia. Point-of-care testing of HbA1c appears to be a suitable approach to timely and accurately revealing diabetes mellitus, and it demonstrates a better quality of diagnosis compared to the standard determination of glycemia. Glycated hemoglobin (hemoglobin A1c, HbA1c, A1C, or Hb1c; also known as HbA1c or HGBA1c) is measured primarily to identify the average plasma glucose concentration over prolonged periods (Tables 21.1 and 21.2).

Table 21.1 ADA criteria for the diagnosis of diabetes and prediabetes based on HbA1c

Status	HbA1c (%)	HbA1c (mmol/mol)
[a]Prediabetes	5.7~6.4	39~47
[b]Diabetes	≥ 6.5	≥ 48

[a]suggested by the Diabetes Control and Complications Trial (DCCT);
[b]suggested by the IFCC

Clinical Biochemistry: A Laboratory Guide
Rooma Devi, Aman Chauhan, Simmi Kharb, and Chandra Shekhar Pundir

ISBN 978-981-4968-75-1 (Hardcover), 978-1-003-45566-0 (eBook)
www.jennystanford.com

Table 21.2 General principle of HbA1c assay in the presence of standard hemoglobin

Hemoglobin, Glycated Hemoglobin		
Different retention in matrix due to surface polarity	Different affinity of antibodies	Different weight of fragments
↓	↓	↓
Chromatography	Immunoassay	Mass spectrometry

Glycated Hemoglobin and Other Advanced Glycation End-Products

HbA1c is a glucose-modified hemoglobin created during the spontaneous reaction between glucose and N-terminal valine residues on β chains of hemoglobin-creating β-N1-deoxy fructosyl.

Measured HbA1c

There are three major HbA1c testing methods currently available to clinical laboratories.

1. Chromatography-based HPLC assay
2. Antibody-based immunoassay
3. Enzyme-based enzymatic assay

Boronate Affinity Chromatography

Boronate affinity chromatography is a glycation-specific method based on boronate binding to the unique cis–diol configuration formed by stable glucose attachments to Hb. The linearity range for the HbA1c detection is 5.3% to 17%.

Latex-Enhanced Immunoassay Method

The latex-enhanced immunoassay for HbA1c is based on the interactions between antigen molecules (HbA1c) and the HbA1c-specific antibodies coated on latex beads.

Enzymatic HbA1c Assay Method

Recent innovations have yielded a Direct Enzymatic HbA1c Assay™, which uses a single channel test and reports %HbA1c values directly, without the need for a separate THb test or a calculation step.

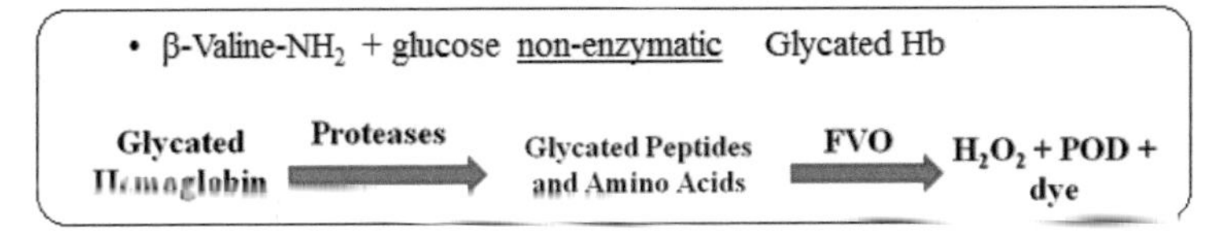

Figure 21.1 Glycated hemoglobin detection based on fructosyl valine oxidase (FVO) enzyme.

Capillary electrophoresis: Basically, two possibilities exist for the separation of HbA1c in capillary electrophoresis (CE) according to the charge-to-mass ratio.

Clinical Significance of Glycated Hemoglobin

- **Increase in red cell turnover:** Blood loss, hemolysis, haemoglobinopathies and red cell disorders, myelodysplastic disease.
- **Interference with the test (this depends on the method used):** Persistent fetal hemoglobin and hemoglobin variants, carbamylated hemoglobin (uremic patients).
- In patients who fluctuate between very high and very low levels, glycated hemoglobin readings can be misleading (the clinician should compare with extra information obtained from home capillary blood glucose tests).
- HbA1c can be useful in identifying patients who may be presenting unrealistically good results of home glucose tests.

Glycated Albumin

- Albumin makes up 60% of all proteins in serum with a concentration of 30–50 g/L.
- Albumin's molecular weight is 66.7 kDa, and it is composed of a single polypeptide chain with 585 amino acids and 17 disulfide chains.
- It also has 24 sites for the formation of AGEs, and glycation occurs by non-enzymatic means.
- Lysine-NH_2 + Glucose $\xrightarrow{\text{non-enzymatic}}$ Glycated albumin
- **Glycation depends on**
  a. Lifespan of RBCs
  b. Blood glucose concentration

- It is present in the whole body.
- Duration of glycemic monitoring and reflects average glucose levels over 2–3 weeks.

Advantages

- Monitoring the effect of changes the therapy
- Anemia
- Low within person variability
- Iron deficiency
- Hemoglobinopathies
- Dialysis
- Pregnancy

Limitations

- Thyroid disease
- Nephrosis
- Thyroid disease

Fructosamine

Fructosamine of serum protein
Location: Whole body
Duration of glycemia reflected: Intermediate (2–3 weeks)

Advantages

- Monitoring the effect of changes in therapy
- Anemia
- Low within person variability
- Iron deficiency
- Hemoglobinopathies
- Dialysis
- Pregnancy

Limitations

- Not been standardized
- Affected by serum protein concentration and proportion
- Thyroid disease
- Nephrosis

Chapter 22

Enzymes

Enzymes are biocatalysts involved in various chemical reactions in the biological system. They are mainly composed of globular proteins and are produced by living cells. They have become increasingly important these days due to their widespread acceptance in the industry, chemical laboratories, dairy products, and other disciplines. They are also a replacement for traditional inorganic catalysts because they are more efficient and versatile. The most desirable character observed in enzymes is that unlike other catalysts in aqueous solutions, they can catalyze a reaction in mild conditions at normal temperature and pressure with very high reaction specificity, bond specificity, and optical specificity. Based on the specificity, enzymes are divided into six major classes.

1. Oxidoreductases: e.g., alcohol dehydrogenase, lactate dehydrogenase
2. Transferases: e.g., alanine alpha ketoglutarate transferase (SGPT, SGOT)
3. Hydrolases: e.g., amylase, pepsin, trypsin
4. Lyases: e.g., fumarase
5. Isomerases: e.g., L-alanine, isomerase
6. Ligases: e.g., glutamine synthetase

Clinical Biochemistry: A Laboratory Guide
Rooma Devi, Aman Chauhan, Simmi Kharb, and Chandra Shekhar Pundir

ISBN 978-981-4968-75-1 (Hardcover), 978-1-003-45566-0 (eBook)
www.jennystanford.com

Enzyme Kinetics

The enzyme xanthine oxidase (XOD) catalyzes the following reaction:

$$\text{Xanthine} + O_2 + H_2O \xrightarrow{\text{XOD}} \text{Uric acid} + H_2O_2$$

The rate of formation of uric acid from xanthine is determined by measuring the increase in the absorbance of uric acid at 290 nm. The xanthine assay is modified based on the oxidation of xanthine to uric acid by xanthine oxidase, performed in vitro. The test mixture contains 1.8 ml of 50 mM sodium phosphate buffer (pH 7.5), 0.1 ml of xanthine (0.16 mM), and 0.1 ml of XOD (150 U/mg). The increase in absorbance at 290 nm is read by a UV spectrophotometer against a blank. The activity is calculated as follows:

$$\text{Units/ml} = \frac{\Delta A/\text{min} \times 1000 \times 3\ \text{ml} \times \text{dilution}}{1.22 \times 10^4 \times 0.1\ \text{ml}}$$

Total reaction mixture volume = 3 ml,
Extinction coefficient uric acid = 1.22×10^4 cm^{-1},
Volume of enzyme = 0.1 ml,
Dilution factor = df.

* 1 unit converts 1.0 nmol xanthine per minute into uric acid. Per ml at pH 7.5 at 25°C.

Factors Affecting Enzyme Activity

Various kinetic properties of XOD enzymes have been studied: effect of pH, incubation temperature for maximum activity, incubation time, and effect of substrate (xanthine) concentration on optimal working conditions.

Effect of pH: To determine the optimal pH for the immobilized enzyme/biosensor response, the pH of the reaction buffer was changed from 5.5 to 8.0 using the following buffers. To determine the optimal pH for the immobilized enzyme/biosensor response, the pH of the reaction buffer was changed from 5.5 to 8.0 using the sodium phosphate buffers (0.05 M). The increase in current (in mA) was measured using a spectrophotometer.

Effect of incubation temperature: The reaction mixture was incubated at different temperatures ranging from 25°C to 50°C to determine the incubation temperature for maximum activity. The enzymatic reaction was measured at these temperatures as described above.

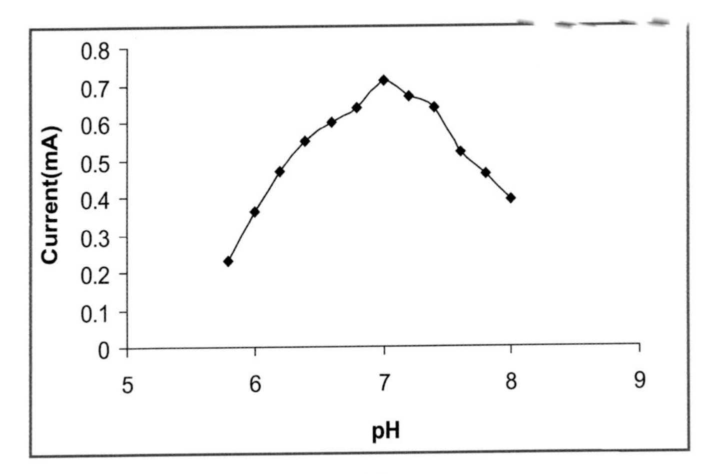

Figure 22.1 Effect of pH (optimum pH 7).

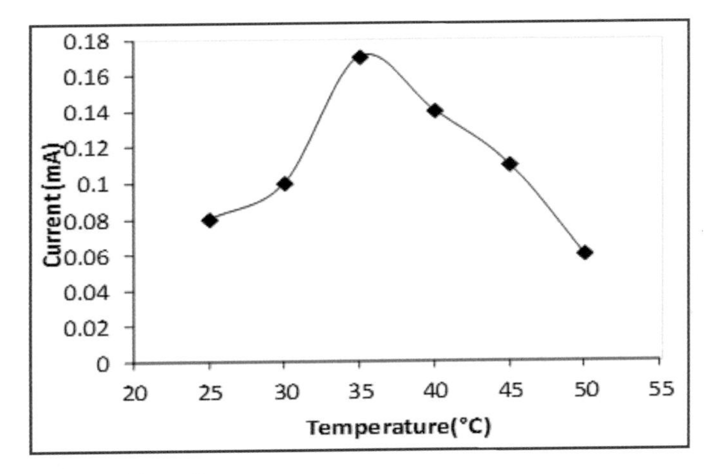

Figure 22.2 Effect of temperature.

Effect of incubation time: The time of maximum activity of the immobilized enzyme reaction was determined by incubating the reaction mixture at various times ranging from 1 to 10 seconds. The enzymatic reaction of absorbance was measured enzymatically at 1 min intervals using a spectrophotometer as described above.

Effect of substrate (xanthine) concentration: The effect of substrate concentration on the initial rate of the immobilized enzyme reaction was studied by varying the final concentration of xanthine in the range of 0.8 μM to 200 μM. The enzymatic reaction was measured by XOD using a spectrophotometer as described above.

Determination of Apparent K_m and I_{max}

We plotted the Lineweaver–Burk graph between the reciprocal of the biosensor response (1/[I]) and the xanthine concentration (1/[S]) and calculated K_m and I_{max} from the plot using the Michaelis–Menten equation.

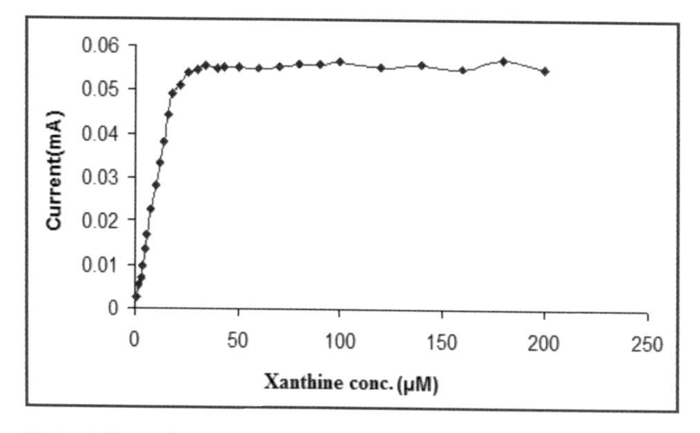

Figure 22.3 Effect of substrate concentration (xanthine).

$$\frac{I}{V} = \frac{K_m}{V_{max}}\left(\frac{I}{S}\right) + \frac{I}{V_{max}}$$

where slope = $\frac{K_m}{V_{max}}$; intercept = $\frac{I}{V_{max}}$

Immobilization of Enzymes

Immobilization means physical confinement or localization of enzyme molecules. This can be done through (a) the physical adsorption of enzymes onto water-insoluble organic or inorganic supports, (b) the entrapment of enzymes within gel matrices or semipermeable microcapsules, and (c) the covalent attachment of enzymes to the water-insoluble supports.

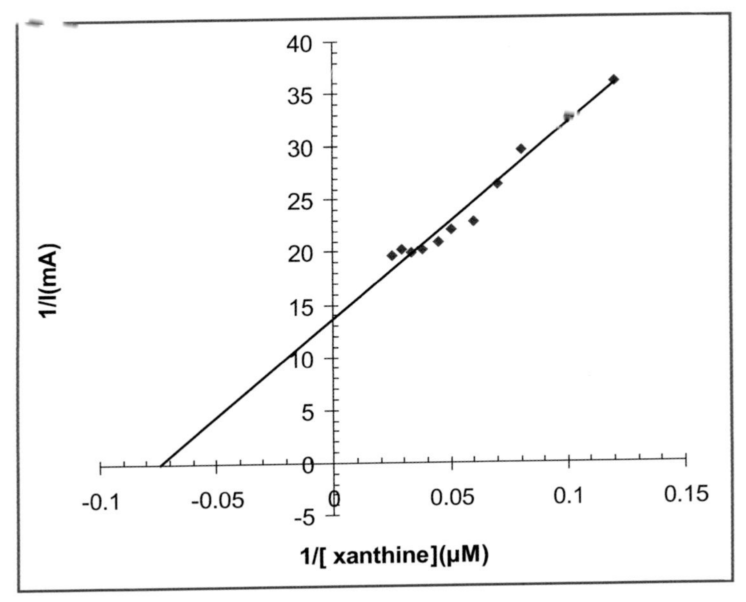

Figure 22.4 Lineweaver–Burk plot for the effect of xanthine concentration.

There are three major reasons for immobilizing enzymes: They (a) offer a considerable operational advantage over freely mobile enzymes, such as reusability, applicability in repeated continuous operations, rapid termination of reaction, and controlled product formation; (b) alter chemical or physical property; and (c) serve as model enzymes or systems for natural, in vivo, and membrane-bound enzymes.

In addition, they are highly stable and resistant to shear stress and contamination. The rate of reaction is fast due to high catalyst concentration, and easy separation of the biocatalyst from the termination medium is possible. Moreover, repeated use of the biocatalyst is also possible. Immobilized enzymes have also certain demerits. The existence of mass transfer resistance and cumbersome immobilization processes are some aspects that can be highlighted against the immobilization of enzymes.

Enzyme Nanoparticles

- Enzyme nanoparticles (ENPs) are used in the construction of improved biosensors.

- Covalent linking was used to make enzyme nanoparticles (ENPs) of size <100 nm, but enzyme particles formed by desolvation are often in the range of 100–200 nm.
- These ENPs exhibit exceptional optical, electronic, electric, thermal, chemical, mechanical, and catalytic properties while supporting a large surface area, compared to the bulk material but less than those of the free enzymes. These properties of ENPs improve the activity of enzyme-based biosensors.
- Direct immobilization of proteins or enzymes onto NPs may cause denaturation, which could result in enzyme inactivation. This problem has been resolved by preparing ENPs and self-crosslinking before their attachment.
- Biosensors based on ENPs have enhanced analytical parameters such as detection limit and current response.

Preparation of Enzyme Nanoparticles by Different Methods

Nanoparticles of soluble proteins have been prepared by their aggregation through the following methods:

- Emulsification in plant oil
- Desolvation by ethanol or natural salts and crosslinking the aggregated protein by glutaraldehyde
- Coacervation via anhydrous ethanol, glutaraldehyde, and ethanolamine
- Crosslinking in water and oil emulsion at high pressure

Characterization of ENPs

ENPs have characteristic features, which can be determined by various techniques as described below.

- **Transmission Electron Microscopy**
- **Colorimetric Methods**

ENPs exhibit characteristic colors. For instance, ENPs of HRP are white in color. The ENP aggregates acquire the color due to a combination of light scattering and absorption. The particles are too large, on the order of the wavelengths of visible light UV absorption spectra and FTIR spectra.

Kinetic Properties of ENPs

- Optimum pH
- Optimum temperature and thermostability
- Response time
- K_m values
- Working range
- Stability and reusability
- Storage stability

Applications of ENPs

Biosensors are quantitative analytical devices consisting of a biological recognition entity such as an enzyme, antibody, phage, aptamer, or single-stranded DNA coupled with appropriate physicochemical, optical, thermometric, piezoelectric, and magnetic transducers. In other words, biosensors are chemical sensors in which the biological recognition element uses a biochemical mechanism interfacing the optoelectronic system.

There are various types of biosensors. Among them electrochemical biosensors have been used widely for the analysis of processed foods and have been given more attention in various fields, due to their small size, cost effectiveness, fast response time, possibility of achieving low detection limits, and good functioning. Some biosensors are as follows:

- Biosensors based on immobilized GOD nanoparticles
- Biosensors based on immobilized ChO_x nanoparticles
- Biosensors based on immobilized uricase nanoparticles
- HbNPs-based biosensors

Multiple Choice Questions

1. The synthesis of prostaglandins from arachidonate is catalyzed by
   a. Cyclooxygenase
   b. Lipoxygenase
   c. Thromboxane synthase
   d. Isomerase

2. Gaucher's disease occurs due to the deficiency of the enzyme
   a. α-Fucosidase
   b. β-Galactosidase
   c. β-Glucosidase
   d. Sphingomyelinase
3. The compound which has the lowest density is
   a. Chylomicron
   b. β-Lipoprotein
   c. α-Lipoprotein
   d. Pre-β-Lipoprotein
4. An example of ligase is
   a. Succinate thiokinase
   b. Alanine racemase
   c. Fumarase
   d. Aldolase

Chapter 23

DNA Isolation from Blood and Tissue

DNA is a genetic material and the basic hereditary unit of each and every cell.

DNA isolation: It is the process of purification of DNA by using a combination of physical and chemical methods. DNA was first isolated in 1869 by Friedrich Miescher. The purification of nucleic acids broadly involves the following steps:

1. Breaking or opening of cells to expose nucleic acids.
2. Separation of nucleic acids from other cellular components.
3. Recovery of nucleic acids in pure form.

Significance of DNA Isolation

- DNA isolation is needed for genetic analysis in scientific, medical, and forensic fields.
- The technique of DNA isolation should lead to good-quality DNA, which is pure and devoid of contaminations such as RNA and proteins.

Principle

The nucleic acid absorbs light strongly in the UV region at 260 nm due to the conjugated bond present in purine and pyrimidine bases.

Clinical Biochemistry: A Laboratory Guide
Rooma Devi, Aman Chauhan, Simmi Kharb, and Chandra Shekhar Pundir

ISBN 978-981-4968-75-1 (Hardcover), 978-1-003-45566-0 (eBook)
www.jennystanford.com

Samples

The sources of DNA isolation include the following:

- Whole blood, hair, bones, nails, tissues, etc.

Instruments

- DNA concentration is measured using a double cell UV spectrophotometer.
- Centrifuge
- Vortex mixer
- Micropipettes

Reagents Used and Their Functions

Reagent	Function
Triton	Triton acts as a detergent and breaks cell membranes.
Sodium dodecyl sulphate (SDS)	SDS causes lysis of the nuclear membrane releasing the DNA within.
Saturated NaCl	NaCl stabilizes DNA and helps in its precipitation.
Chloroform	Chloroform denatures proteins and keeps DNA in the aqueous phase.
Isopropanol	Isopropanol precipitates DNA.
Tris EDTA (ethylene diamine tetra acetate) buffer, nuclease free water	It protects the nucleic acid from degrading by DNase or RNase.

Procedure

It basically consists of the following steps:

1. **Preparation of cell extract**

   To extract DNA from tissue/cell of interest, the cells have to be separated, and cell membranes have to be extracted by an extraction buffer.

   - Ethylene diamine tetra acetate (EDTA) removes Mg^{2+} ions that preserve the structure of cell membranes.

- Sodium dodecyl sulfate (SDS) disrupts the cell membrane by removing the lipids from the membrane.
- After the centrifugation, partially digested organelles leave the cell extract as a clear supernatant.

2. **Purification of DNA from Cell Extract**

   DNA is located inside the nucleus; therefore, all other biological components must be removed from the cell, including all protein and cell debris, etc. For purification, the most commonly used procedures are as follows:

   - Ethanol precipitation
   - Phenol chloroform extraction
   - Minicolumn purification

   Cellular and histone protein bound to the DNA can be removed either by adding proteases and precipitated with sodium or ammonium acetate extraction of protein is done by phenol-chloroform mixture.

3. **Concentration of DNA samples**
   - DNA is concentrated by mostly ethanol precipitation. In the presence of salt, ethanol efficiently precipitates polymeric nucleic acid.

4. **Measurement of purity of DNA concentration**
   - DNA concentration can be measured by UV absorbance spectrochemistry.
   - Absorbance is measured at 260 nm wavelength. An absorbance of 1.0 corresponds to 50 µg of dsDNA per ml.
   - $A_{260}:A_{280}$ (A260 nm and 280 nm ratio) = 1.8 implies pure DNA sample.
   - $A_{260}:A_{280} < 1.8$ implies that the DNA is contaminated with either protein or phenol.

Procedure

Step 1	Blood/tissue sample (0.5 ml) ↓
Step 2	Cell lysis buffer triton (1 ml) ↓ Centrifuge at 3000 rpm for 2 min

Step 3	SDS nuclei lysis buffer (0.3 ml) ↓ Incubate at room temperature for 2 min
Step 4	Saturated NaCl (0.1 ml) + Chloroform (0.6 ml) ↓ Centrifuge at 6000 rpm for 2 min
Step 5	Transfer 0.3–0.5 ml of supernatant to a new microfuge tube. ↓
Step 6	Add cold isopropanol (0.6 ml) ↓ Centrifuge at 13,000 rpm for 1 min
Step 7	Discard the supernatant and dry at room temperature ↓ Add 0.05–0.1 ml nuclease-free water.
Step 8	Centrifuge for mixing, and store at −20°C.

Precautions

- All glassware, plasticware, and reagents must be autoclaved before use.
- Phenol is corrosive, so it should be handled carefully.

Chapter 24

Mass Spectrometry

Mass spectrometry has become an attractive molecular histology tool in pharmaceutical and medical research. It is a widely used instrumental technique in physics, analytical chemistry, and forensic laboratories. It is based on the ionization and fragmentation of sample molecules in the gas phase.

After the ionization of the analytes of interest into various fragments, the spectrum generated is interpreted based on mass-to-charge ratios of the different ions. An important part of this technique is sample preparation prior to mass analysis.

It consists of four major components:

1. Sample inlet
2. **Ion source:** It desorbes and ionizes the analyte.
3. **Mass analyzer:** Here ions are separated based on their mass-to-charge ratios (m/z).
4. **Detector:** It detects the separated ions.

Desorption ionization: In this type of ionization, the sample to be analyzed is first dissolved with a matrix and then placed in the path of high-intensity photons. This is called matrix-assisted laser desorption. Why are we using a matrix here? A matrix is used to keep analytes protected from direct laser beams.

Clinical Biochemistry: A Laboratory Guide
Rooma Devi, Aman Chauhan, Simmi Kharb, and Chandra Shekhar Pundir

ISBN 978-981-4968-75-1 (Hardcover), 978-1-003-45566-0 (eBook)
www.jennystanford.com

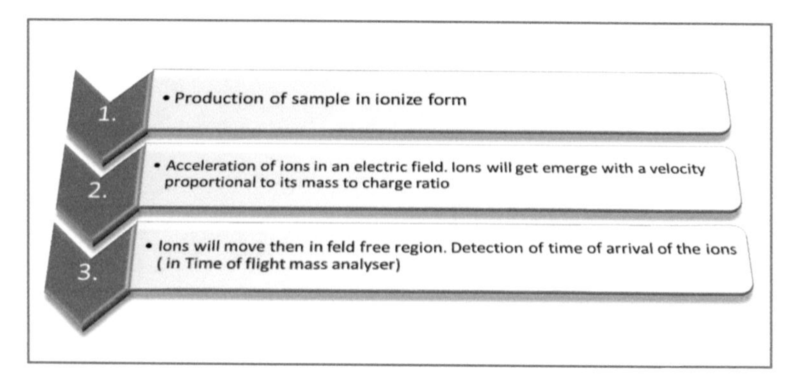

Figure 24.1 Flow chart of various steps in mass spectrometry.

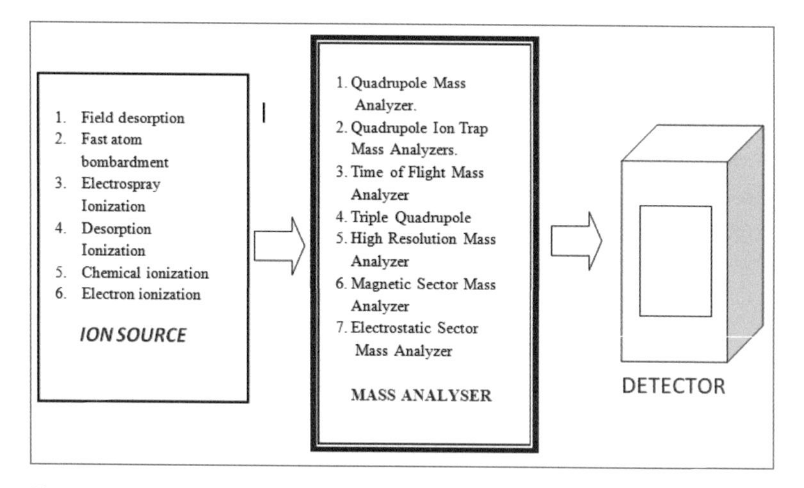

Figure 24.2 Overview of mass spectrometry.

Electrospray ionization: In this type of ionization, a narrow stainless steel capillary is used. The sample is first dissolved in a polar solvent and then passed through this capillary. A high voltage of 3 or 4 kV is then applied on the tip of the capillary. Due to the electric field, the sample coming out of the tip gets dispersed into charged droplets. Further due to solvent evaporation, analyte ions get separated from the solvent.

Electron ionization: The work of ionization is done with the help of high-energy electrons.

Types of Mass Analyzers

1. Quadrupole Mass Spectrometer

It is the most common mass spectrometer in use today. It consists of four cylindrical rods, which are placed parallel to each other. The electric field on the two sets of diagonally opposed rods is responsible for selecting sample ions based on their mass-to-charge ratio (m/z) and allowing only ions of selected m/z value to pass through the detector. The other ions (unselected ions) get deflected from the rods. Quadrupole mass spectrometers are used to determine the mass of molecules that are 4000 Da or less.

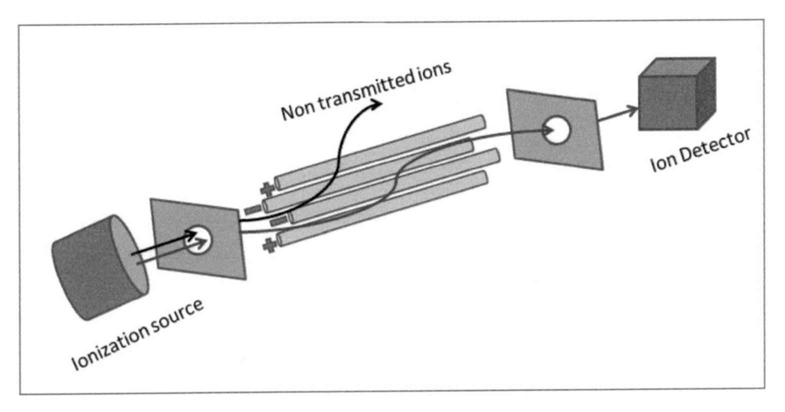

Figure 24.3 Diagrammatic presentation of quadrupole mass spectrometers.

Concept of Full Scan and Selected Ion Monitoring

This mass scan can be as wide as in the full scan analysis or as narrow as in selected ion monitoring. A full scan gives a qualitative picture of the composition of the sample. Full scan mass spectrometry is used for qualitative applications to obtain structural information. On the other hand, selected ion monitoring (SIM) is a mass spectrometry scanning mode in which only a limited mass-to-charge ratio range is transmitted/detected by the instrument, as opposed to the full spectrum range. Thus, SIM is used for target compound analysis.

1. **Ion trap mass analyzer**

   Quadrupole ion trap mass analyzers work on the same physical principles as quadrupole mass analyzers. The only difference is that in this type, the ions are trapped in stable orbits inside a three-dimensional chamber. These trapped ions are then sequentially ejected based on their m/z. A quadrupole ion trap mass spectrometer consists of the following:

a. Three hyperbolic electrodes: a donut-shaped ring electrode
b. An entrance endcap electrode
c. An exit endcap electrode

It involves accumulation and destabilization of ions. Ions travel through a hole present in each endcap electrode. The ring electrode is located midway between the two endcap electrodes. The ion trap consists of a circular ring electrode and two endcaps that form a "trap" (instead of four parallel rods, which are present in quadrupole mass spectrometers). First, ions get trapped and then the electric field is adjusted to destabilize these ions.

2. **A triple quadrupole mass spectrometer or triple quad**

It is a tandem mass spectrometer consisting of two quadrupole mass analyzers in series, with a (non-mass-resolving) radio frequency (RF)-only quadrupole between them to act as a cell for collision-induced dissociation. This whole configuration is abbreviated as $Q_1q_2Q_3$.

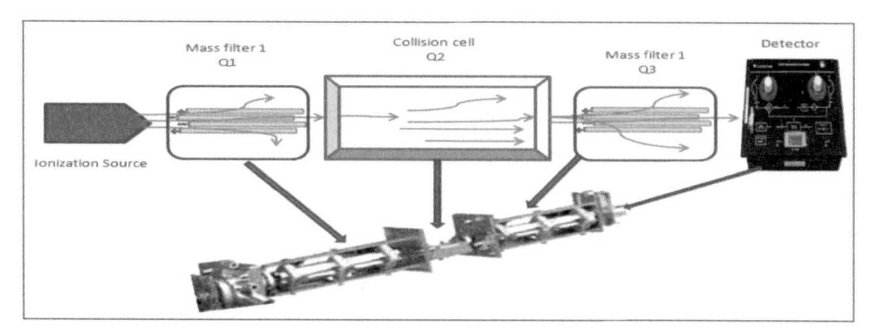

Figure 24.4 Diagrammatic presentation of a triple quadrupole mass spectrometer.

Q_1 (mass filter): Select an ion of interest (precursor ion).

Q_2 (collision cell): The ion of interest will move into collision cell and get fragmented due to collision between ion and gas molecules.

Q_3 (mass filter): To analyze the product ions generated in Q_2.

What is a precursor ion and a product ion?

- **Precursor ion:** A single ion that passes through the first analyzer.
- **Product ion:** The ions formed during fragmentation.

3. **Time-of-flight mass spectrometer**

   It is a high-resolution mass spectrometer. Unlike a quadrupole mass spectrometer, a time-of-flight (TOF) mass spectrometer is a pulsed and non-scanning mass spectrometer consisting of an accelerator, a field-free region, a reflector, and a flight tube. It detects ions of different m/z by calculating the time taken for the ions to travel through a field-free region. The ions generated in an ionization unit get accumulated and introduced to a flight tube in pulses. They are further accelerated when a high voltage is applied between the electrodes. Each ion flies at its unique velocity inside the flight tube to reach the ion detector, which depends upon the molecular mass. So, the velocity will be higher for ions with smaller masses and lower for ions with larger masses. It can measure a large number of analytes simultaneously and is useful in drug-screening applications. A TOF mass spectrometer is used to determine large masses of complete proteins that are more than 4 KDa.

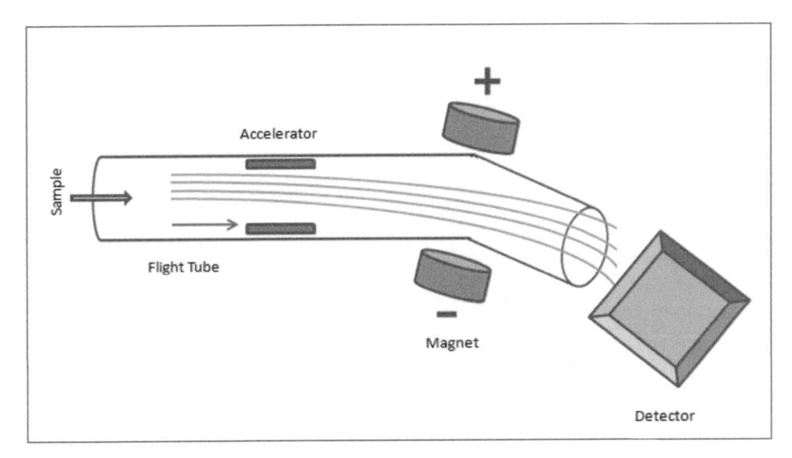

Figure 24.5 Flow diagram of a time-of-flight (TOF) mass spectrometer.

4. **Orbitrap mass spectrometer**
   It is another high resolution mass spectrometer. It is an ion trap mass analyzer consisting of an outer barrel-like electrode and a coaxial inner spindle-like electrode that traps ions in an orbital motion around the spindle. An orbitrap mass analyzer is an extremely powerful ion trap instrument that allows the system to act as both a mass analyzer and a detector. It can achieve 100,000 to 250,000 FWHM.

Mass spectrometry is coupled with liquid chromatography or gas chromatography, which makes it a powerful technique. Liquid chromatography–mass spectrometry (LC–MS) is an analytical chemistry technique that combines the physical separation capabilities of liquid chromatography (or HPLC) with the mass analysis capabilities of mass spectrometry. Coupled chromatography–mass spectrometry systems are popular in chemical analysis because the individual capabilities of each technique get enhanced. While liquid chromatography separates mixtures with multiple components, mass spectrometry provides structural identity of the individual components with high molecular specificity and detection sensitivity.

Detectors Used in Mass Spectrometer

1. **Electron multipliers (EM):** Dynodes are used for the amplification of current of ions. Dynodes (series of dynodes) are arranged in such a manner that when ions strike with them, then the electrons are produced which strike with other dynodes and so on. Finally current is measured as voltage pulse.

Figure 24.6 Presentation of amplification by dynodes.

2. **Photomultiplier conversion dynode:** The ions initially strike a dynode, which results in the emission of electrons. These electrons then strike a phosphor screen, which results in the release of photons. Amplification further occurs in a cascade fashion when photons pass into the multiplier.
3. Array detectors
4. Faraday cup (FC)

Applications of Mass Spectrometry

1. **Biochemical screening for genetic disorders:** Used for screening inborn errors of metabolism (IEMs) (screening of neonatal dried blood spot). Approximately 20 IEMs can be detected using this technique, for example, aminoacidemia (phenylketonuria, maple syrup urine diseases, tyrosinemia, etc.), organic acidemias (propionic acidemias, methylmalonic academia, etc.).
2. **Therapeutic drug monitoring and toxicology:** LC–MS assays have been developed for monitoring immunosuppressants such as cyclosporin, tacrolimus, sirolimus, everolimus, and mycophenolic acid. In the LC–MS assays, several drugs and metabolites can be measured in one run, which is a useful feature of these assays. It detects a wide range of drugs, toxins, and their metabolites.
3. **Measurement of steroid hormone and vitamin D:** LC–MS assays are also useful for the estimation of vitamin D and steroid hormones.
4. **Forensic sample analysis:** In forensic analysis, it is a very useful method as the quantity of sample present is minute. Thus, a highly sensitive method is required in such cases. LC–MS is used in such conditions.
5. **Pharmaceutical science:** LC–MS is used for the evaluation of impurities during drug preparations.
6. **Proteomics:** It is used in the identification, quantification, and post-translational modification analysis of protein structure.
7. **Mass spectrometry imaging (MSI):** It is a technology used to visualize the spatial distribution of molecules.

Chapter 25

Flow Cytometry

Flow cytometry is a technique used to detect, identify, and count specific cells. This method can also identify specific components within cells. So, it is used for the detection and measurement of physical and chemical characteristics of cells or particles in a heterogeneous fluid mixture. Lasers as light sources are used to produce both scattered and fluorescent light signals, which are then read by detectors such as photodiodes or photomultiplier (PMT) tubes.

The three basic components of a flow cytometer (traditional flow cytometer) are as follows:

1. **Fluidic system:** The sample is injected into a stream of sheath fluid, which is usually a buffered saline solution within the flow chamber. This flow chamber then focuses the sample at the center of the sheath fluid where the laser beam then interacts with the particles (term used is hydrodynamic focusing).

   So, this component is meant for transporting particles from the fluid stream to the laser beam for interrogation.
2. **Optic system:** It consists of the laser and lenses (used to shape and focus the laser beam to the flow of the sample). During the interrogation of particles with laser, light will get reflected or diffracted. The factors that affect light scattering are cell complexity, cell membrane, nucleus, and other granular

Clinical Biochemistry: A Laboratory Guide
Rooma Devi, Aman Chauhan, Simmi Kharb, and Chandra Shekhar Pundir

ISBN 978-981-4968-75-1 (Hardcover), 978-1-003-45566-0 (eBook)
www.jennystanford.com

materials inside the cell. The scattered light is measured in two different directions:

a. **The forward direction (forward scatter or FSC):** It indicates the relative size of the cell. So, forward scattering is directly proportional to the cell size.

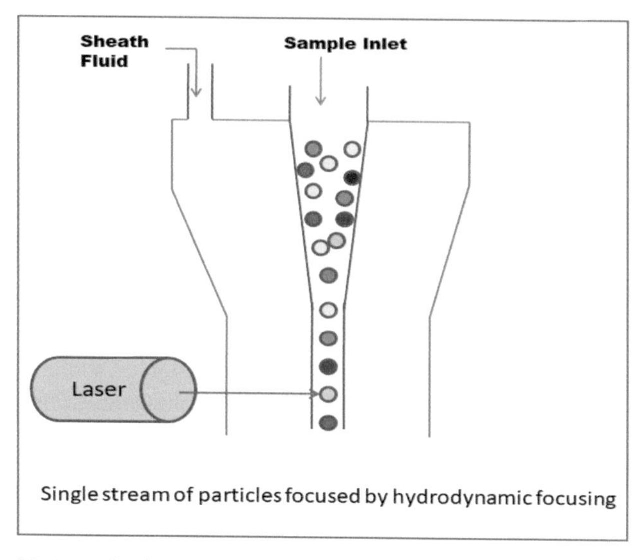

Figure 25.1 Hydrodynamic focusing in flow cytometry.

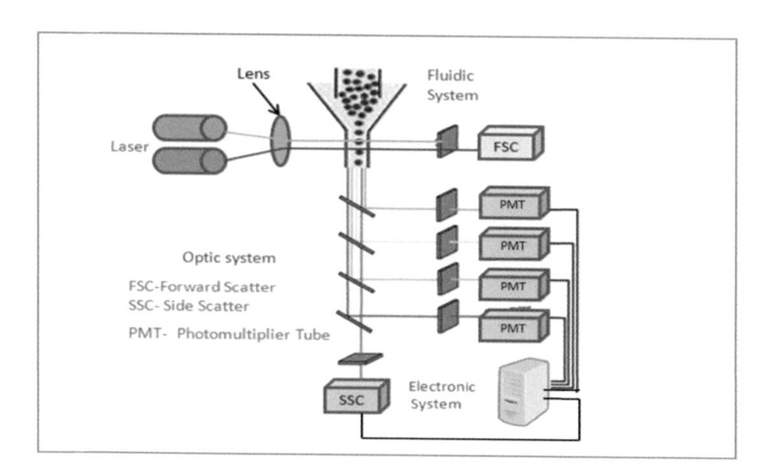

Figure 25.2 Overview of flow cytometry.

b. **At 90° (side scatter or SSC):** It indicates the internal complexity or granularity of the cell. So, side scattering is directly proportional to the granularity of the cell. Dead cells have lower FSC and higher SSC than living cells

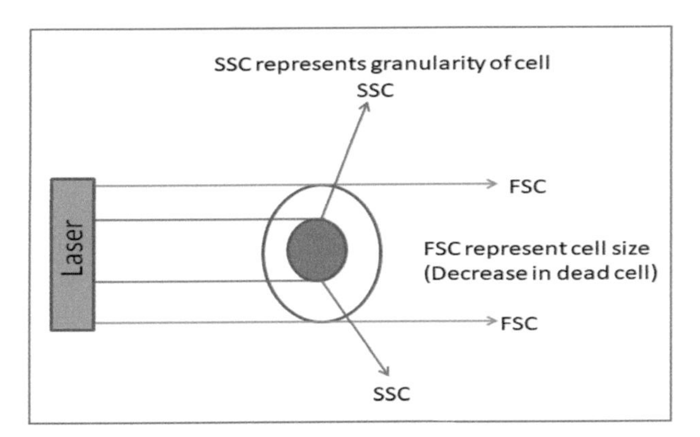

Figure 25.3 Light scattering when laser interrogates the cell.

Detectors perpendicular to the beam are used to measure SSC and fluorescence. Fluorescence is produced when a fluorescent compound absorbs light at a particular wavelength and gets excited to a higher state. When this compound comes back to the ground state, it emits excess energy as a photon of light. This transition of energy is called fluorescence. The most common fluorochromes used in flow cytometry are FITC (fluorescein isothiocyanate) and PE (phycoerythrin).

3. **Electronic system:** When particles pass through laser, light signals are generated. The electronic system converts these signals from the detectors into digital or electronic signals. These electronic signals strike one side of the PMT or the photodiode, which get converted into a relative number of electrons, which are multiplied to create a more significant electrical current. A photodiode is less sensitive to the light signals as compared to PMT.

Types of Cell Sorters

- **Quartz cuvette:** It has fixed laser alignment.
- **Jet in air:** It requires the lasers to be aligned daily and is more difficult to set up. However, it is more suitable for small particle detection.

Some other types of flow cytometry are as follows:

- **Acoustic focusing cytometer:** It uses ultrasonic waves to better focus cells for laser interrogation.
- **Imaging cytometer (IFC):** It combines traditional flow cytometry with fluorescence microscopy.
- **Mass cytometer:** It combines time-of-flight mass spectrometry and flow cytometry. In this type, cells are labeled with heavy metal ion-tagged antibodies and detected using time-of-flight mass spectrometry.

Flow cytometry data are typically reported in two distinct ways: a histogram and/or a dot plot.

Uses

1. **Apoptosis**
   - The phospholipids in the plasma membrane are asymmetrically distributed in live cells (phosphatidylserine present in the plasma membrane of leaflets). Annexin V (calcium-dependent phospholipid-binding protein) can easily bind to the negatively charged phosphatidylserine. In live cells, as phosphatidylserine is present in the inner leaflet, annexin V will not be able to bind it. However, in the case of apoptosis due to the loss of asymmetry, the phosphatidylserine gets exposed to the outer leaflet and annexin V binds to phosphatidylserine. So, apoptotic cells can be detected based on the increased binding of annexin V.
   - Detection by propidium iodide (PI) and 7-amino actinomycin D (7-AMD):

     The uptake of PI dye by apoptotic cells is another method for the detection of apoptosis. However, the uptake of PI by apoptotic cells is much lower than that of necrotic cells. We can summarize it as follow:

No uptake	Live cells
PI-Dim	Apoptotic cells
PI-Bright	Necrotic cells

2. **Identification of Platelet Disorders**

   Platelets have two receptors on their surface: Gp1b (CD42)

and Gp2b/3a (CD41/CD61). For the normal functioning of platelets, these two receptors must be present on the surface. Two main disorders can be easily identified by flow cytometry:

a. **Bernard–Soulier syndrome:** It is a platelet adhesion defect (absence of CD42).
b. **Glanzmann thrombocytopenia:** It is a platelet aggregation disorder (absence of CD41/CD61).

3. **Plasma Cell Disorders**

With the help of different cell markers, we can easily identify normal cells and neoplastic cells. Reactive B cells have both kappa and lambda chains, but malignant B cells have either kappa or lambda chains.

Normal Cells (Markers Present)	Neoplastic Cells (Markers Present)
CD38, CD138, CD19, CD45 (present)	• CD38, CD138 (present) • Cytoplasmic light chain (present) • CD19, CD45, CD56 (negative) • Surface immunoglobulin (negative)

4. **Paroxysmal Nocturnal Hemoglobinuria (PNH)**

CD55 (DAF) and CD59 (MIRL) markers are absent in PNH, which can be easily analyzed by flow cytometry.

5. **Leukemias**

Multiparametric flow cytometry provides rapid and accurate determination of antigen expression profiles in leukemias. By detecting different types of markers, we can easily differentiate between different types of leukemias.

ALL	AML	CLL
1. CD10, CD19, CD34, TdT, and HLA-DR (positive) 2. CD45 (negative) 3. Surface kappa or lambda chain (negative)	CD13, CD33, CD117, CD34, TdT, and HLA-DR (positive)	CD5 and CD23 (positive)

6. **Cell Sorting or Fluorescence-Activated Cell Sorting (FACS)**

It allows the physical separation of a cell or particle of interest from a heterogeneous population. A mixture of fluorescently stained cells is passed through a small aperture. A fluorescent-labeled antibody specific to a particular cell surface protein is added to the mixture of cells. The fluorescent character of interest of each cell is measured as soon as it gets passed through laser. The stream of liquid holding different cells will break into individual droplets containing different cells due to a vibrating mechanism (vibrating nozzle). At this break-off point, the droplets containing different cells get charged. The fluorescence detector identifies the signal from each drop containing different cells, and it applies an electric charge to these drops. These drops then pass through an electrostatic deflection system. The positively charged drops get deflected toward the negative electrode, and the negatively charged drops get deflected toward the positive electrode. Some drops holding cells will not get charged, and they will not get deflected toward either side. Based on their charges, they will be collected in different tubes.

7. **Cell Cycle Study**

   It is used to know the distribution of cells in the three major phases of cell cycle (G1, S, and G2/M). The most common method to assess the cell cycle by flow cytometry is to measure the DNA cell contents.

   a. G0 phase—beginning of the cell cycle
   b. G1 phase—beginning of cell growth
   c. S phase—nuclear DNA content doubles to 4N
   d. G2 phase—cell growth maintained here to 4N
   e. M phase—phase of actual cell division

To measure the DNA content, the cell must be stained with different dyes and their staining is directly proportional to the amount of DNA content.

	Dye	Place of Binding
1.	DAPI	Binds to the A–T rich region
2.	Chromomycin A3	Binds to the G–C rich region
3.	Bromodeoxyuridine	Incorporated into the genome during the S phase and detected using anti-BrdU antibodies
4.	Hoechst 33342	Binds to the minor groove of dsDNA
5.	Propidium iodide	Intercalating agent, which stains the cellular genome upon fixation of cell

Interpretation of Results of Flow Cytometry

Flow cytometry data are represented in two ways:

1. **Histograms:** They measure or compare only a single parameter. In histograms, the intensity of a single channel is plotted on the *x*-axis and the number of detected events on the *y*-axis.
2. **Dot plots:** They compare two or three parameters simultaneously, e.g., intensity of side scatter versus forward scatter.

Multiple Choice Questions

1. Which of the following is larger in size?

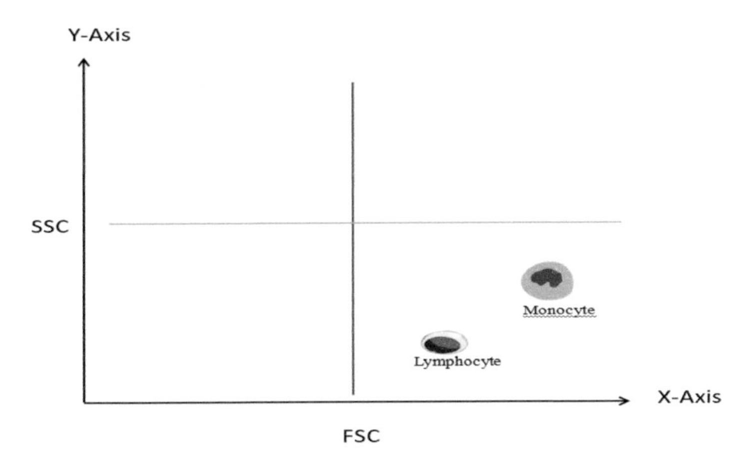

   a. Monocytes
   b. Lymphocytes
   c. Both have the same size
   d. None of the above

   Ans: a

   As we have discussed earlier, size is directly proportional to forward scattering. The more the forward scattering, the larger the size.

2. The uptake of PI by apoptotic cells (PI-Bright) represents
   a. Live cells

b. Necrotic cells
c. Apoptotic cells
d. All of the above

Ans: b

The uptake of PI by apoptotic cells is much lower than that by necrotic cells. We can easily differentiate this by the following table:

No uptake	Live cells
PI-Dim	Apoptotic cells
PI-Bright	Necrotic cells

3. Which of the following is correct in terms of granularity?

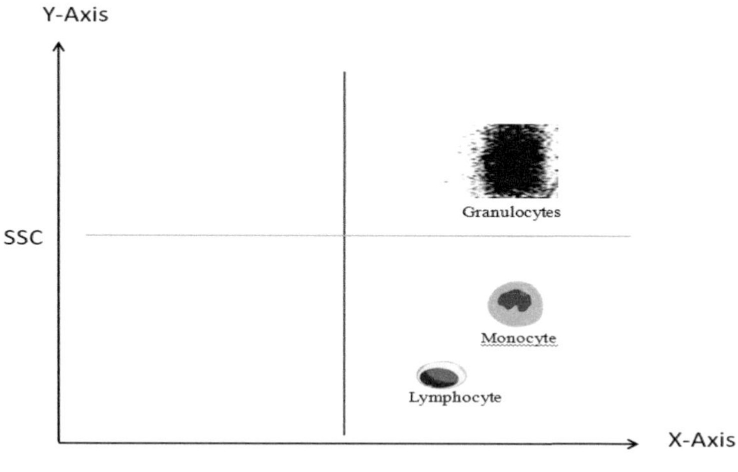

a. Monocyte > Granulocyte > Lymphocyte
b. Lymphocyte > Monocyte > Granulocyte
c. Granulocyte > Monocyte > Lymphocyte
d. None of the above

Ans: c

Side scatter is directly proportional to granularity. On the *y*-axis, the cell with more side scattering will be more granular.

4. Hydrodynamic focusing is used in which of the following:
   a. Flow cytometry

b. FRET
c. Centrifugation
d. None of the above

Ans: a

The flow chamber in flow cytometry focuses the sample to the center of the sheath fluid, where the laser beam then interacts with the particles.

5. Identify the condition by interpreting the dot plot.

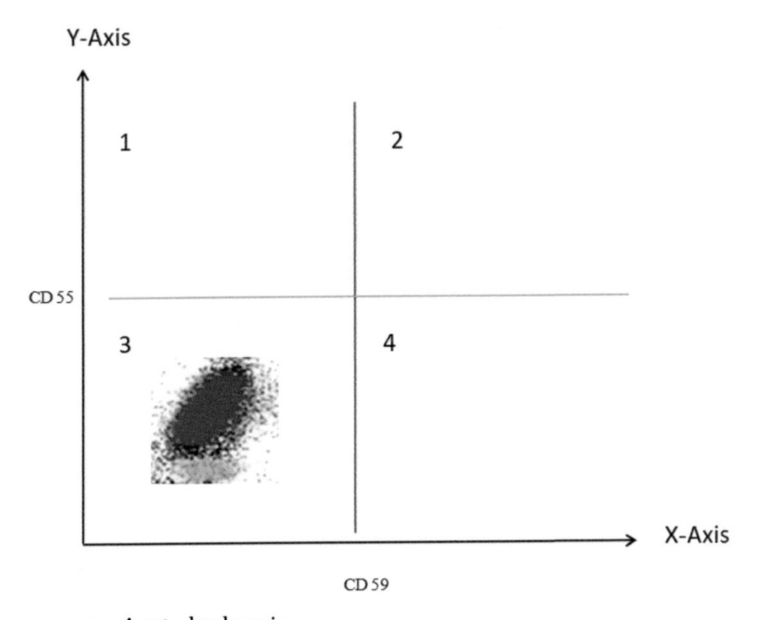

a. Acute leukemia
b. PNH
c. Bernard–Soulier disease
d. Glanzmann thrombocytopenia

Ans: b

CD55 (represented on the *x*-axis) and CD59 (represented on the *y*-axis) are markers used for identifying PNH. PNH can be identified if CD55 and CD59 are not found in flow cytometry (quadrant 3, i.e., double negative).

6. Identify the condition shown in the following plot:

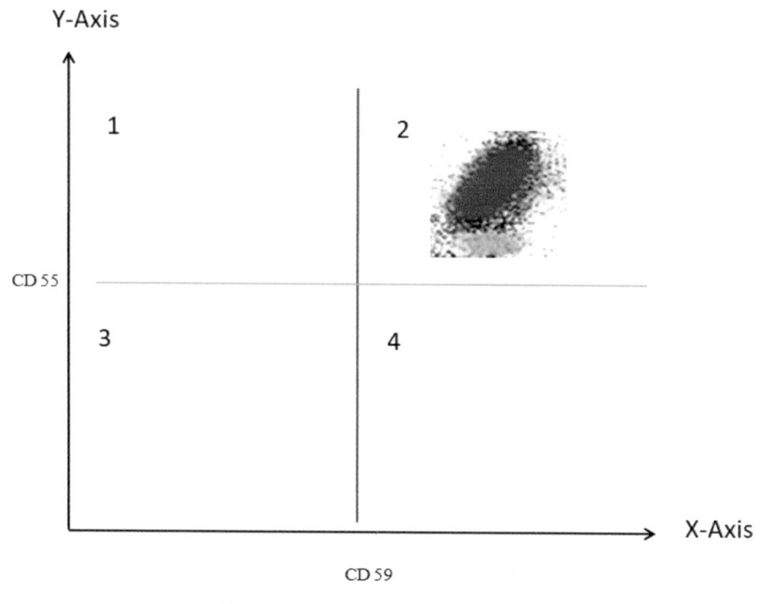

a. Acute leukemia
b. PNH
c. Bernard–Soulier disease
d. Normal cells

Ans: d

CD55 (represented on the *x*-axis) and CD59 (represented on the *y*-axis) are markers used for identifying PNH. PNH can be identified if CD55 and CD59 are not found in flow cytometry (quadrant 3, i.e., double negative). But in this condition, both markers are present (quadrant 2, i.e., double positive).

Chapter 26

Centrifugation

Centrifugation is the process of using centrifugal force to separate lighter components of a solution, mixture, or suspension from heavier components. A centrifuge is a device that separates various components of a solution based on shape, size, density, viscosity of the medium, and rotor speed. It works on the principle of centrifugal force.

Basic Theory of Sedimentation

A body moving in a circular motion with an angular velocity omega (ω) is subjected to a centrifugal force (f) acting in the outward direction. This throws the heavier particles in the outside direction.

The centrifugal force depends on the radius of revolution and the angular velocity (in rpm):

$$f = \omega^2 r$$

The centrifuge works on the principle of gravity and generation of a centrifugal force to sediment different fractions. The rate of sedimentation depends on the applied centrifugal field (f) directed radially outward. Here, f depends on the angular velocity ω (rad/s) and the radial distance r (cm) of the particle from the axis of rotation.

Clinical Biochemistry: A Laboratory Guide
Rooma Devi, Aman Chauhan, Simmi Kharb, and Chandra Shekhar Pundir

ISBN 978-981-4968-75-1 (Hardcover), 978-1-003-45566-0 (eBook)
www.jennystanford.com

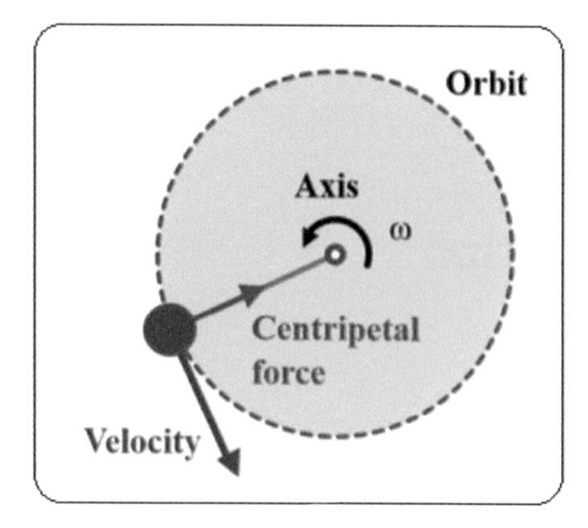

Figure 26.1 Basic theory of sedimentation.

Rotation around a single axis creates a rotational force that goes in the direction of a circle.

The accelerating force that is created shoots outward perpendicular to the circle from all directions.

Sedimentation depends on

- The size of particles
- The density difference between particles and the medium
- The radial distance from the axis of rotation to liquid meniscus (r_t)
- The radial distance from the axis of rotation to the bottom of the tube (r_b)

The centrifuge works on the sedimentation principle, where the centripetal acceleration causes denser substances and particles to move outward in the radial direction. The greater the difference in density, the faster they move. If there is no difference in density (isopycnic conditions), the particles stay steady.

Types of Centrifuges

1. **Bench-top/table-top centrifuges**
   - These are ordinary centrifuges.
   - Their speed ranges from 2000 to 8000 rpm.
   - They are used for the separation of precipitate serum and suspended particles in urine or other body fluids.

2. **High-speed centrifuges**
   - They can attain a speed of 10,000 to 20,000 rpm.
   - They are used in blood banks, research laboratories, and pharmaceutical industry.
   - They are mostly floor top.
   - They can be used at even cold temperatures or at lower than the room temperature. They called refrigerator centrifuges.
3. **Ultracentrifuges**
   - They can attain speeds of 70,000 to 90,000 rpm.
   - They are mostly used for polymer compounds.

Type of Centrifuge Based on Rotors

- Fixed angle rotors
- Swinging bucket rotors/horizontal rotors
- Vertical rotors

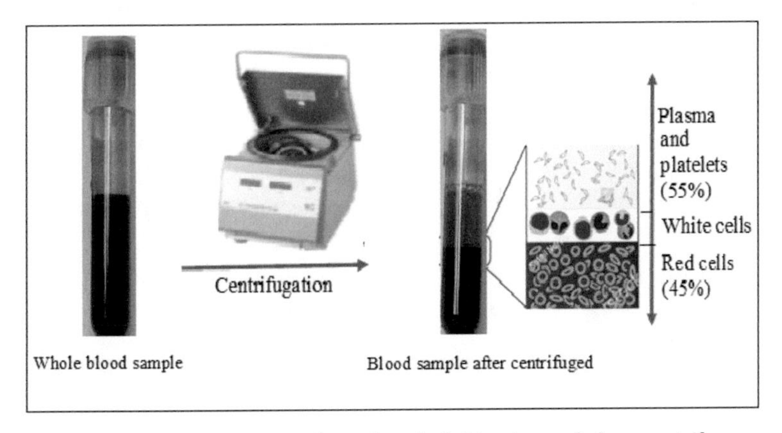

Figure 26.2 Serum separates from the whole blood sample by a centrifuge.

Types of Centrifugation and Their Applications

- Move boundary/zonal centrifugation
- Rate-zonal/velocity-gradient centrifugation
- Density-gradient centrifugation
- Rate-zonal/density-gradient centrifugation
- Isopycnic centrifugation

Centrifugation Techniques	Applications
Preparative	
Differential velocity (moving boundary) centrifugation (pelleting)	Separation and isolation of particles in a solution. May be applied to precipitates, cell organelles, cells, or biomolecules.
Fractional centrifugation	Isolation of particles, based on size, by successive centrifugation at increasing rotor speeds. May be used to separate cell organelles.
Analytical	
Differential centrifugation	Sedimentation of particles in a medium of homogeneous density. Used to measure the sedimentation coefficient, size, and molecular weight of a particle.
Sedimentation equilibrium centrifugation	Used to determine the molecular weight of a macromolecule or other particle.
Density-gradient centrifugation	
Rate-zonal density-gradient centrifugation	A gradient is present in the tube before centrifugation, and the sample is layered on top. Used to isolate purified molecules and determine *s*.
Isopycnic centrifugation	A gradient is formed during centrifugation. Used to isolate purified molecules and determine *s*.

Precautions

- The centrifuge tube should never be fully filled.
- The cup or tubes of the centrifuge should be balanced properly.
- Before placing the centrifuge, the metal bucket and holes should be checked for any debris such as broken glass.
- Place the centrifuge on a firm base.
- Never open the chamber until the rotor has come to a complete stop.

Applications

- Separation of serum or plasma from RBCs.
- Separation of sediment in urine.
- Separation of a protein-free filtrate.
- Washing of RBCs by normal saline.
- Separation of Ag-bound fraction or Ab-bound fraction in immunoassays.

Biochemistry OSPE

Spot 1

1. Identify the equipment.
2. Mention its uses.
3. What is the principle of centrifugation?
4. What are the other equipment used in biochemistry laboratories?
5. What is the recommended speed?
6. Write its mechanism of action.

Question

1. Separate serum from a blood sample with the help of a centrifuge.

Chapter 27

Automation

Laboratory automation is a complex integration of robotics, computers, liquid handling, and some other technologies. It has dramatically changed both analytical and non-analytical aspects of clinical laboratory operations. It involves the use of various control systems for operating clinical laboratory equipment and other applications with minimum human intervention, thus reducing the possibility of human error. Automation enables anyone to perform many tests with the help of analytical instruments without an analyst. By using automation, one can get low cost per test, do more tests in a short period of time, improve accuracy, reduce sample or reagent consumption, improve reproducibility, and minimize variations in results.

An automated analyzer is a medical laboratory instrument used to measure biological samples quickly. Auto-analyzers can be divided into two types:

1. Open system
2. Closed system

Clinical Biochemistry: A Laboratory Guide
Rooma Devi, Aman Chauhan, Simmi Kharb, and Chandra Shekhar Pundir

ISBN 978-981-4968-75-1 (Hardcover), 978-1-003-45566-0 (eBook)
www.jennystanford.com

Open System	Closed System
The auto-analyzer is assembled by pulling together different parts. Users can buy reagents from any company to reduce the cost.	Users have to buy reagents from the manufacturer only as the reagents from other providers are not compatible with the auto-analyzer and the machine will not run in this case.

What is a modular system and an integrated system?

Like a modular kitchen, each part of a modular system is designed in such a way that it easily fits in different systems. If there is a problem in any part of the system, it can be replaced without affecting the entire system. But in an integrated system, the entire machine is designed in such a way that every part is integrated with the machine. So, services and maintenance become expensive and difficult in this system.

What is batch analysis and random-access analysis?

Batch analysis is beneficial if one parameter is to be analyzed on multiple samples, e.g., blood sugar in multiple specimens. In random-access analysis, multiple tests can be simultaneously performed on one sample.

Batch Analysis	Random-Access Analysis
Analysis of a large number of specimens in one run; can be used by continuous flow, centrifugal analysis, and discrete analysis.	Only offered by discrete analyzers, which have the capability to run multiple tests on one sample at the same time or run one test on multiple samples at the same time.

Automatic analytical systems are of the following types:

A. Continuous Flow Analyzer

- It can be used for frequently requested independent analysis, such as blood glucose and blood total protein.
- It is a single-channel/continuous flow/batch analyzer.
- Samples, reagents, and diluents flow through a small bore tubing continuously in a sequential manner, and air bubbles are introduced at precisely defined intervals, forming unique reaction segments.

- Sequential analysis means samples are tested one after the other and results are reported in the same order.
- These air bubbles, which were introduced at regular intervals, serve as separating and cleaning media as well as prevent carry-over effects.
- Samples, reagents, and diluents are injected into the flowing carrier solution.
- The sample gets mixed with the diluents and reagents, which are further sent through tubing and mixing coils.
- The sample and reagents are mixed with the help of coiled tubing.
- A dialyzer is used for separating or filtering interference substances.
- An oil-heating bath is used to promote color development or the completion of enzymatic reaction.
- The sampler probe must be placed in distilled water to avoid blockages and precipitation (when there is no sample to be analyzed).
- More sophisticated continuous flow analyzers are also available, which use parallel single channels to run multiple tests on each sample.
- Detection is done by measuring the absorbance with the help of a spectrophotometer through a continuous flow cell at certain wavelengths.
- One of its drawbacks is that it consumes more reagents to maintain the flow, thus increasing the cost per test for users.

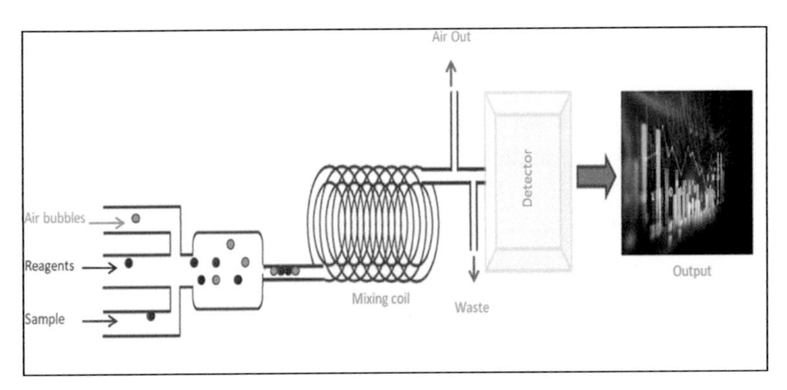

Figure 27.1 Flow diagram of a continuous flow analyzer.

B. Discrete Analyzer

- Discrete analyzers are the most popular and versatile analyzers and have almost completely replaced continuous flow and centrifugal analyzers.
- It is non-continuous type of analyzer and it has random-access fluid (hydrofluorocarbon liquid) which reduces surface tension between samples/reagents and their tubing, therefore, it has no chance of carry over (sample contamination).
- It has the capability of running multiple tests on one sample at the same time or one test on multiple samples at the same time. So multiple tests can be run on the sample in a run.
- Reagents used are in the liquid form (wet chemistry analyzer) or the dry lyophilized form/slides (dry chemistry analyzer).
- The sample and reagents are mixed by spinning the cuvette at a high speed followed by a sudden stop or by introducing the reagent into the cuvette through jet action (the sample and reagents are mixed by spreading the sample layer over the reagent layer).
- No automated methodology is used to remove interfering substances from the reaction mixture, so methods having minimum effects should be used.
- Each analysis, even for the same analyte or sample, takes place at different cups, and this is the main principle of discrete processing.
- It is also known as a random-access analyzer as it is selective and performs only on those assays that were ordered on each sample.
- The system analyzes only the tests programmed as per requirement, e.g.,
  - Sample 1: Urea, creatinine, uric acid
  - Sample 2: TAG, HDL, total cholesterol
  - Sample 3: Total protein, albumin
- Discrete-type analyzers use kinetics instead of endpoint methods, so they give more accurate results.

C. Centrifugal Analyzer

- In centrifugal analyzer, only single test can be done at one time, so it is a type of batch analyzers in which analysis is sequential, discrete, and parallel.
- Centrifugal force is used for transferring and mixing the samples and reagents. When the rotor holding the cuvette is spun at a high speed, the reagents and samples get mixed. The start–stop sequence of rotation or bubbling of air through the samples and reagents is used for mixing.
- The rotor moves the final product up to the optical system for the final reading.
- The analyzer has a sample compartment and a reagent compartment in the cuvette. The samples and reagents are dispensed by the auto-sampler, and the reagent probe into their respective compartments.
- **Drawback:** Only one test can be performed at a time. Centrifugal analysis uses the motion of a spinning rotor for mixing, thermal equilibration, transport, and measurement.
- Sample and reagents are pipetted into different chambers in a rotor.

Major Steps in Automated Analysis

1. Specimen Preparation and Identification

- This remains a manual process in most laboratories and includes the clotting time (if serum is used). Centrifugation and the transfer of the sample to the analyzer cup (unless primary tube sampling is used) cause delays and expenses in the testing process. However, some laboratories use robotics to minimize human errors at this step, for example Abbott-Vision.
- An important aspect in sample preparation and identification is that the sample must be properly identified, which can be achieved by labeling the sample cups and using labeled analysis positions in the auto-analyzer.

- Nowadays, the most sophisticated approach is the use of barcodes on the collection tubes. These barcodes have information related to patients' personal information and all tests.

2. **Specimen Measurement and Delivery**
   - The slots in the trays or racks are usually numbered to get rid of sample identification errors.
   - These trays or racks move automatically in one-position steps at particular speeds, which decide the number of specimens to be analyzed per hour.
   - When the discrete instrument is in operation, the probe automatically dips into each sample cup and aspirates a portion of the liquid.

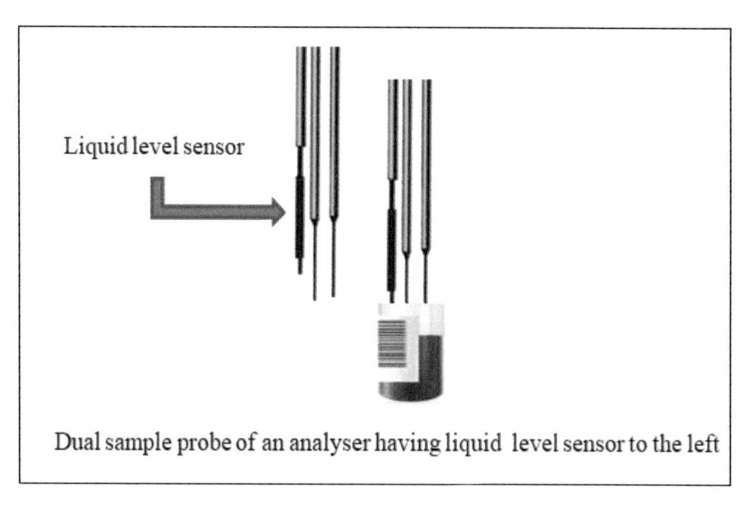

Figure 27.2 Dual sample probe of analyzer.

   - Sampling probes on instruments programmed in such a way that it reaches only to a prescribed depth in the sample cup to aspirate the maximum amount of sample. So sample probe will not touch the surface of the sample cup and prevents its damage.
   - The analyzers also have a parallel liquid-level-sensing probe, which controls the entry of the sampling probe to a minimum depth below the surface of the serum, which

avoids clogging of the probe with serum separator gel or clot.

- In continuous flow analyzers, samples as well as reagents are pumped continuously and bubbles are pumped at definite interval to form the reaction chamber.
- In the VITROS analyzer (dry chemistry analyzer), the four quadrants fit on a tray carrier (sample cup trays are quadrant).
- Each quadrant holds 10 samples (all four quadrants can hold 40 samples).

3. **Reagent Systems and Delivery**
   - Reagents may be classified as liquid (used in wet chemistry analyzers) or dry (used in dry chemistry analyzers).
   - Dry reagents may be in the form of lyophilized powder (reconstituted diluent or buffer) or a multilayered dry chemistry slide for the VITROS analyzer. (In multilayered dry chemistry analyzers, cartridges for different parameters are used, which hold the slides.)
   - To deliver reagents, many discrete analyzers use techniques similar to those used to measure and deliver the samples. Syringes, pressurized reagent bottles, or piston pumps driven by a stepping motor pipette the reagents into reaction containers.
   - In VITROS analyzers, multiple layers on the slide are backed by a clear polyester support.
   - The VITROS analyzers work on the principle of reflectance photometry.
   - In VITROS analyzers, there is no requirement of daily calibration as well as sample pretreatment as they have a filter layer for this purpose. On the other hand, there is high accuracy and speed and also a small amount of sample is required although the dry chemistry reagents are difficult to store and have high cost.
   - The slides in VITROS analyzers contain:
     - A spreading layer
     - Scavenger layer/filter layer
     - An indicator layer, where the analyte of interest may be quantified

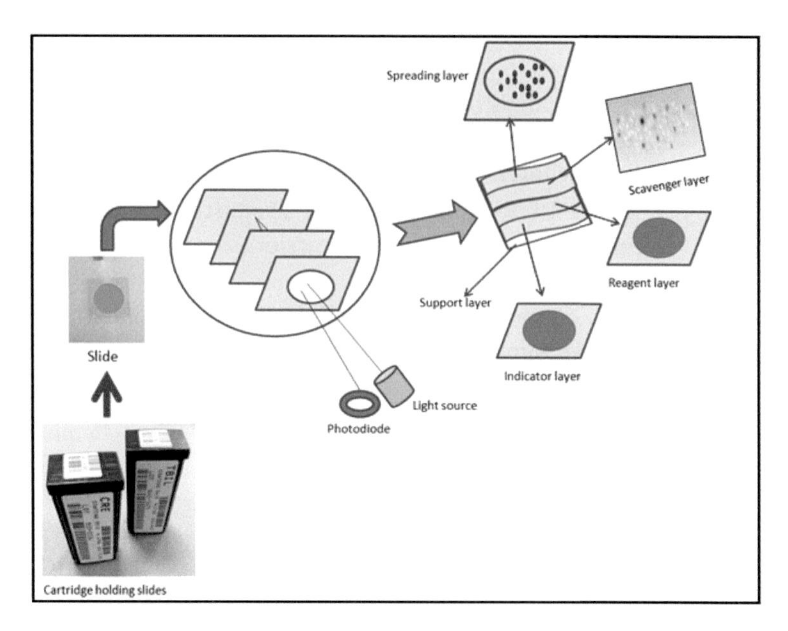

Figure 27.3 Various layers in dry chemistry analyzers and principle of detection.

- Indicator layer
- In VITROS analyzers, a precondition station is used to bring the temperature of each slide close to 37°C before it enters the incubator.

4. **Chemical Reaction Phase**

   This phase includes:
   - Mixing
   - Separation
   - Incubation
   - Reaction time

Chapter 28

Cerebrospinal Fluid Examination

Cerebrospinal fluid (CSF) is formed by the selective dialysis of plasma by the choroid plexus present in the ventricles of the brain. CSF is present in the cavity that surrounds the brain in the skull and the spinal cord in the spinal column. The total volume of CSF is about 150 ml. CSF diagnoses central nervous system (CNS) abnormalities.

Functions of CSF

- It helps to protect the brain and the spinal cord from injury by acting like a buffer.
- It also acts as a medium for the transfer of substances from the brain tissue and the spinal cord to blood.
- It provides buoyancy and nutrition to the brain.
- It removes waste from the brain.
- It serves as a medium for nutrient exchange in the CNS.

Sample Collection

Cerebrospinal fluid is collected via lumbar puncture (LP) usually done between the third and fourth lumber vertebra under aseptic conditions. It is collected in sterile vials for biochemical and bacteriological examinations.

Clinical Biochemistry: A Laboratory Guide
Rooma Devi, Aman Chauhan, Simmi Kharb, and Chandra Shekhar Pundir

ISBN 978-981-4968-75-1 (Hardcover), 978-1-003-45566-0 (eBook)
www.jennystanford.com

The fluid is collected in three different vacuette tubes:

1. **For biochemical tests:** use fluoride (gray cap) plain vacuette
2. **For cell count:** use EDTA vacuette (lavender cap)
3. **For bacterial culture:** use plain sterile vacuette (green cap)

Sample Stability

- Analyze immediately after collection.

Composition of CSF

Glucose	45–80 mg/dl
Proteins	15–45 mg/dl
Sodium (Na^+)	147 meq/L or increase
Chloride	120–130 meq/L or increase
Calcium (Ca^{2+})	2.3 meq/dl or decrease
Urea	12.0 mg/100 ml
Creatinine	1.5 mg/100 ml
Lactic acid	18.0 mg/100 ml
Bilirubin	Absent

Characteristics of CSF

Nature	
Color	Clear, transparent fluid
Specific gravity	1.004–1.007
Reaction	Alkaline, does not coagulate
Cells	
Adults	0–5 cells/cumm
Infants	0–30 cells/cumm
1–4 years	0–20 cells/cumm
5–18 years	0–10 cells/cumm

Indications of CSF Examination

- **Infections:** Meningitis, encephalitis

- **Inflammatory condition:** Neurosyphilis, sarcoidosis, SLE
- **Infiltrative condition:** Leukemia, lymphoma

Clinical Significance

Cerebrospinal fluid is tested in disorders such as:

- Tuberculosis
- Meningitis, multiple sclerosis
- Brain tumor, spinal cord tumor
- Subarachnoid hemorrhage

Routine

- Gross examination
- Cell count
- Glucose
- Protein

CSF Profile

Condition	Glucose	Proteins	Cells
Bacterial Meningitis	↓↓↓	⬆⬆	⬆ Neutrophiles
TB	↓	⬆	⬆ Lymphocytes
Fungi	—	⬆	Monocytes
Viruses	—	⬆	Lymphocytes
Autoimmune	—		Lymphocytes

Biomarkers

- SB100
- TAU protein
- Amyloid beta

Chapter 29

Oxalic Acid

Oxalate is a metabolic breakdown product of the Krebs cycle in eukaryotes and the glyoxylate cycle in other microorganisms.

- Normal range of oxalic acid
  Urine: 10–30 mg/24 h
  Plasma: 0.8–2.50 μmol/L
- Oxalate concentration is higher in whole blood than in serum.
- Ammonium, lithium oxalate, sodium, and potassium inhibit blood coagulation by forming an insoluble complex with Ca^{2+}.
- Oxalate oxidase is of great importance in the measurement and diagnosis of primary and secondary hyperoxaluria.
- Hyperoxaluria leads to the formation of calcium urinary oxalate stones, malabsorption, steatorrhea, ethynylene glycol poisoning, and ileal disease.
- Crystals of oxalate are sharp and can be large enough to irritate the body, which leads to the formation of kidney stones.
- A decreased level of oxalate in urine is also associated with hyperglycosuria and hyperglycinemia.
- Oxalate is extracted from urine and subsequently assayed by measuring the amount of hydrogen peroxide produced in an oxidation reaction catalyzed by oxalate oxidase.

Clinical Biochemistry: A Laboratory Guide
Rooma Devi, Aman Chauhan, Simmi Kharb, and Chandra Shekhar Pundir

ISBN 978-981-4968-75-1 (Hardcover), 978-1-003-45566-0 (eBook)
www.jennystanford.com

Principle

Enzymatic method: Both the enzymes, oxalate oxidase and oxalate decarboxylase, have been used for the measurement of oxalate.

Oxalate decarboxylase: This enzyme acts on oxalate ion by degrading it into formate and carbon dioxide, which involves the direct measurement of the product (Fig. 29.1).

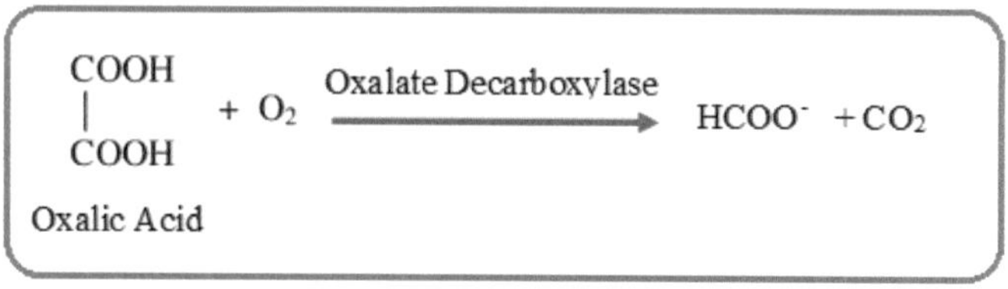

$$\underset{\text{Oxalic Acid}}{\begin{matrix} COOH \\ | \\ COOH \end{matrix}} + O_2 \xrightarrow{\text{Oxalate Decarboxylase}} HCOO^- + CO_2$$

Figure 29.1 Reaction mechanism of oxalate decarboxylase.

Urinary oxalate can be determined by a simple and rapid enzymatic method using oxalate decarboxylase. CO_2 generated from decarboxylated oxalate is released into alkaline buffer, and the change in pH is measured.

Oxalate oxidase: This enzyme acts on oxalate to form CO_2 and H_2O_2 in the presence of oxygen according to the following reaction.

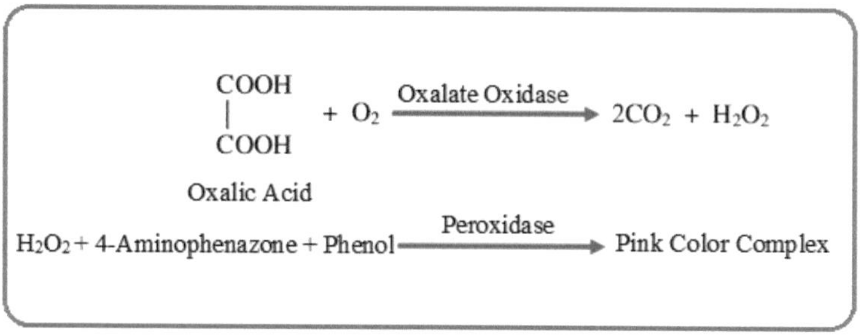

$$\underset{\text{Oxalic Acid}}{\begin{matrix} COOH \\ | \\ COOH \end{matrix}} + O_2 \xrightarrow{\text{Oxalate Oxidase}} 2CO_2 + H_2O_2$$

$$H_2O_2 + \text{4-Aminophenazone} + \text{Phenol} \xrightarrow{\text{Peroxidase}} \text{Pink Color Complex}$$

Figure 29.2 Reactions of oxalate oxidase assay.

H_2O_2 is measured by a colored reaction, which is directly proportional to oxalate concentration. The oxalate is determined by using various techniques as mentioned earlier. The enzyme exhibits maximum activity at pH 5.0 and 40°C. The rate of H_2O_2 formation is linear up to 2 min.

Oxalate Decarboxylase Assay

Method

- The concentration of oxalate is measured by a conventional Warburg manometric technique.
- One unit of activity releases 1 µmol of CO_2 per minute at optimum working conditions (37°C and pH 3.0).

Reagents

- Potassium oxalate (0.1M) at pH 3.0
- Potassium citrate (1.0 M) at pH 3.0

Enzyme

Enzyme concentration: 3–5 units/ml

Procedure

The following reagents are used:

1.0 M Potassium citrate, pH 3.0	0.4 ml
Reagent grade water	2.4 ml
Diluted enzyme	0.1 ml

- Pipette 0.1 ml of 0.1M potassium oxalate (pH 3.0) into the side arm.
- Include a flask containing no enzyme as a blank, and another flask containing 3.0 ml of water to serve as a thermal barometer.
- After 10 min of equilibration, close the manometers, tip in and mix the substrate, and replace the flasks in the bath.
- Read all the flasks every 5 min for 30 min.

$$\text{Units/mg} = \frac{\text{Microliters of } CO_2 \text{ released per minute}}{22.4 \times \text{mg enzyme in reaction mixture}}$$

Other Methods for Determination of Oxalate

Direct precipitation: The method is very simple, and a quantity as small as 5×10^{-7} mol/L can be measured in urine. However, urine contains many inhibitors such as magnesium, polyphosphate, and other electrolytes.

Solvent extraction: Oxalic acid was extracted by using different solvents such as diethyl ether and tri-n-butyl-phosphate. The method was considered better than the direct precipitation method, but due to low recovery, it did not solve the problem.

Isotachophoresis: The method is simple and rapid but requires specialized hands to handle it and is also expensive.

Ion chromatography: In this method, oxalic acid is separated from interfering compounds. Ion exchange chromatography is preferred when a large number of samples are analyzed.

High-performance liquid chromatography: The method is specific, sensitive, and easy to set up. It is an excellent method for separating low-molecular-weight substances from biological fluids. In this method, oxalate is extracted using tri-n-butyl phosphate.

Chemiluminescent method: This method was developed for the determination of oxalate based on the oxidation of methyl red by dichromate.

Clinical Significance of Oxalates

- Kidney stones
- Calcium oxalate stones occur in about 75–80% of cases.
- Uric acid stones occur in about 5–10% of cases.
- Struvite stones are also called infection stones, urease, or triple phosphate stones.

Hyperoxaluria: It occurs due to the excessive excretion of oxalic acid. Hyperoxaluria is of four types:

- Primary hyperoxaluria
- Enteric hyperoxaluria
- Idiopathic hyperoxaluria
- Oxalate poisoning

Index